THE HYDROPHOBIC EFFECT: FORMATION OF MICELLES AND BIOLOGICAL MEMBRANES

THE HYDROPHOBIC EFFECT: FORMATION OF MICELLES AND BIOLOGICAL MEMBRANES

CHARLES TANFORD

James B. Duke Professor of Physical Biochemistry
Duke University

A WILEY-INTERSCIENCE PUBLICATION

JOHN WILEY & SONS, New York • London • Sydney • Toronto

Library of Congress Cataloging in Publication Data

Tanford, Charles
The hydrophobic effect: formation of micelles and biological membrances

"A Wiley-Interscience publication"
1. Surface Chemistry. 2. Solution (Chemistry) 3. Micelles
4. Membranes (Biology) I. Title.

QP521.T37 547'.1'34514 73-13505
ISBN 0-471-84460-8

Printed in the United States of America

10 9 8 7 6 5 4 3 2

PREFACE

The hydrophobic effect is perhaps the most important single factor in the organization of the constituent molecules of living matter into complex structural entities such as cell membranes and organelles. It is equally important in the formation of detergent micelles and other phenomena that occur in aqueous solution. In spite of this, no comprehensive account of the hydrophobic effect exists, and this book is intended to fill that gap. It begins with the simplest manifestation of the hydrophobic effect—the meager solubility of hydrocarbons in water—and continues with more complex phenomena, such as the formation of micelles, surface films, and lipid bilayers, and the interaction of proteins with hydrophobic ligands. The final chapters relate these topics to the structural and functional properties of biological membranes, which are as yet only poorly understood and the subject of much current research.

In discussing these topics the emphasis is on experimental results and on deductions that can be rigorously derived from them. The underlying thermodynamic principles are presented in some detail, but speculative models that are not directly supported by experimental data are not discussed. A chapter on the structure of water is included, since therein must lie the basis for the hydrophobic effect; here too, however, the focus is on experiment rather than on theoretical models.

I am indebted to numerous colleagues who have helped to clarify my ideas about the subject matter of this book: in particular to Professors J. A. Reynolds and W. Kauzmann, for their valuable suggestions concerning several parts of the manuscript; to Dr. R. Smith, who carried out several technically difficult experimental studies; and to Professor J. D. Robertson, who helped initiate me into the membrane field. I also wish to express my appreciation to the National Science Foundation and to the National Institutes of Health for their support of my research, some of the results of which form an integral part of this book.

CHARLES TANFORD

March 1973
Durham, North Carolina

CONTENTS

"The antipathy of the paraffin-chain for water is, however, frequently misunderstood. There is no question of actual repulsion between individual water molecules and paraffin chains, nor is there any very strong attraction of paraffin chains for one another. There is, however, a very strong attraction of water molecules for one another in comparison with which the paraffin-paraffin or paraffin-water attractions are very slight."

G. S. HARTLEY (1936)

"Possibly the decisive step (in the origin of life) was the formation of the first cell, in which chain molecules of at least two of three types now represented by nucleic acids, proteins, and polysaccharides were enclosed in a semi-permeable membrane which kept them together but let their food in."

J. B. S. HALDANE (1954)

ONE

INTRODUCTION

Life as we know it originated in water and could not exist in its absence. The importance of water's unique physical properties and of its unique solvent power, especially for ions, has been frequently cited (e.g., Henderson, 1913), but the equally great importance of its unique lack of solvent power for many nonpolar substances has received far less attention. That is what this book is about: molecules that are soluble in alcohol, ether, and many other solvents, but not in water; and molecules of a dual nature, with one part soluble in water and another part that is expelled from it. Molecules in the latter category are forced by their duality to adopt unique orientations with respect to the aqueous medium, and to form suitably organized structures. Such molecules are especially crucial to living matter in that they play a prominent role in its organization. The cell membrane, which in effect defines the living cell and allows it to exist as a coherent entity, is perhaps the best example. There can be little doubt that its formation is spontaneous, dependent only on the fact that its constituent molecules are partly hydrophilic and partly hydrophobic.

ORIGIN OF THE HYDROPHOBIC EFFECT

Hydrophobic substances are defined as substances that are readily soluble in many nonpolar solvents, but only sparingly soluble in water, distinct from substances that have generally low solubility in all solvents because they form solids with strong intermolecular cohesion. This distinction is probably especially important from the biological point of view, since it means that molecules expelled from water as a result of their hydrophobicity will tend to remain in a fluid, deformable state.

The existence of hydrophobic substances, and of organic molecules of a dual nature, containing hydrophobic portions, has been known for a long time, but misconceptions about the origin of the phenomenon have persisted until quite recently. Thus McBain, who can probably be credited with prime responsibility for demonstrating the reversible formation of micellar aggregates in aqueous soap solutions (McBain and Salmon, 1920), believed that the association between hydrocarbon chains in the formation of such micelles arises from "like to like" attraction, implying that the attraction of hydrocarbon chains for each other plays an important role in this process (McBain, 1950). Even Debye (1949) evidently shared this belief, since he equated the energy gained in micelle formation with the energy difference between liquid and gaseous hydrocarbon.

In fact the attraction of nonpolar groups (such as hydrocarbon chains) for each other plays only a minor role in the hydrophobic effect. The effect actually arises primarily from the strong attractive forces between water molecules which, being isotropically arranged, must be disrupted or distorted when any solute is dissolved in the water. If the solute is ionic or strongly polar, it can form strong bonds to water molecules, which more than compensate for the disruption or distortion of the bonds existing in pure water; and ionic or polar substances thus tend to be easily soluble in water. No such compensation occurs with nonpolar groups, and their solution in water is accordingly resisted. In terms of the intermolecular bonds that are formed, the process of solution of a nonpolar solute S in water may be represented schematically as

$$2\,H_2O - H_2O + S - S \rightleftharpoons 2\,H_2O - S - H_2O \qquad \text{(I)}$$

and the major factor driving the equilibrium in this reaction to the left is the strength of the hydrogen bonds between water molecules. These bonds are in fact so strong that the schematic representation of the product of reaction I as $H_2O - S - H_2O$ is probably misleading, since the hydrogen bonds tend to be maintained in distorted form, that is, $\overset{\displaystyle\ulcorner\!\!-\!\!-\!\!-\!\!-\!\!-\!\!-\!\!\urcorner}{H_2O - S - H_2O}$ would be a better representation, at least at low temperatures (Frank and Evans, 1945). It is then a loss of entropy rather than bond energy that leads to an unfavorable free energy change for the process.

The true origin of the hydrophobic effect was qualitatively understood, or at least intuitively understood, by Traube (1891), on the basis of the phenomenon of surface activity. Traube showed that molecules consisting of a long hydrocarbon chain attached to a polar group (fatty acids, alcohols, amides, etc.), which have a measurable solubility in water by virtue of their constituent polar groups, tend to migrate to the surface of an aqueous solution, their presence there being recognized by a decrease in the surface tension. At very low concentrations of these molecules in the bulk solution, the reduction in surface tension increases linearly with the bulk concentration, but as the bulk concentration is raised it has progressively less effect. Traube recognized this as a saturation phenomenon, anticipating the more formal expression of this phenomenon by the adsorption equation of Langmuir (1917). The important aspect of the results in relation to the origin of the hydrophobic effect is that at low bulk concentrations, where the surface is sparsely populated and contacts between hydrocarbon chains are minimal, the ratio of concentration at the surface layer to that in the bulk medium increases approximately threefold for each CH_2 group added to the alkyl chain ("Traube's rule"). The same result was obtained for several homologous series involving different polar groups and, since attraction between alkyl groups cannot be important here, there is no ambiguity in the interpretation: the cause of the effect must reside solely in a lack of affinity between hydrocarbon and water.

The same conclusion may be reached in another way. If attraction between hydrocarbon chains is important in migration to the surface, the process would become cooperative at low solute concentrations, and the surface concentration would not be linearly proportional to the concentration in bulk.

It should be noted that the correct interpretation of the hydrophobic effect, assigning the predominant role to the properties of water per se, has been given in several places: by Hartley (1936) in an elegant monograph on micelle formation in aqueous solution, by Frank and Evans (1945) in their interpretation of the thermodynamic properties of aqueous solutions of all kinds of hydrophobic or partly hydrophobic substances, and by Kauzmann (1954) in his analysis of the forces stabilizing the native structure of proteins.

REFERENCES

Debye, P. (1949). *Ann. N.Y. Acad. Sci.*, **51,** 575.

Frank, H. S., and M. W. Evans. (1945). *J. Chem. Phys.*, **13,** 507.

Haldane, J. B. S. (1954). *New Biol.*, **16,** 12.

Hartley, G. S. (1936). *Aqueous Solutions of Paraffin-Chain Salts*, Hermann & Cie., Paris.

Henderson, L. J. (1913). *The Fitness of the Environment*, Macmillan Co., New York.

Kauzmann, W. (1959). *Adv. Protein Chem.*, **14,** 1.

Langmuir, I. (1917). *J. Am. Chem. Soc.*, **39,** 1848.

McBain, J. W. (1950). *Colloid Science*, D. C. Heath and Co., Boston.

McBain, J. W., and C. S. Salmon. (1920). *J. Am. Chem. Soc.*, **42,** 426.

Traube, J. (1891). *Ann.*, **265,** 27.

TWO

THE SOLUBILITY OF HYDROCARBONS IN WATER

The simplest hydrophobic substances include inert gases, hydrocarbons, and some other nonpolar organic substances. The amphiphilic substances that form micelles are nearly all substituted hydrocarbons, and the biological lipids that play the major role in the formation of cell membranes all contain aliphatic hydrocarbon chains as their hydrophobic entity. Thus, among simple hydrophobic substances, hydrocarbons are closest to the main theme of this book. This chapter provides a measure of their hydrophobicity, in terms of the free energy of transfer of hydrocarbon molecules from water to a purely hydrocarbon solvent. For straight-chain aliphatic hydrocarbons, this free energy will be seen to be a strictly linear function of the number of CH_2 groups in the chain.

For reasons first given by Gurney (1953) and reiterated by Kauzmann (1959), free energies of transfer from one solvent to another should be expressed in *unitary* units, which simply means that the solute concentration in the equation for the chemical potential is expressed in mole fraction units. Thus, for a hydrocarbon dissolved in water

$$\mu_W = \mu_W^\circ + RT \ln X_W + RT \ln f_W \tag{2-1}$$

where X_W is the concentration of solute in mole fraction units, f_W the activity coefficient at that concentration, and μ_W° the standard chemical potential on the unitary scale. The reference state is the state of infinite dilution *in water*, which means that all interactions of an isolated hydrocarbon molecule with the solvent water are included in the term μ_W°, and $RT \ln f_W$ represents only that part of the excess chemical potential that arises from interactions of solute molecules with each other. Because of the very low solubility of hydrocarbons in water, $RT \ln f_W$ can safely be set equal to zero at actual experimental concentrations.

The advantage of using mole fraction units for the solute concentration lies in the fact that $RT \ln X_W$ is the correct expression (in very dilute solutions) for the *cratic* part of the chemical potential, which is the purely statistical contribution to the chemical potential that arises from the entropy of mixing solvent and solute molecules. The standard potential μ_W° thus contains no contributions from this effect and represents *only* the internal free energy of the solute molecule and the free energy of its interaction with the solvent. If some other concentration units had been used, μ_W° would include the contribution made by the entropy of mixing (on the basis of ideal mixing) at unit concentration, for example, at 1 mole/liter if molar concentrations are used. There would be a similar factor in the standard chemical potential in a hydrocarbon solvent (μ_{HC}°), but since 1 liter of water and 1 liter of liquid hydrocarbon generally contain different numbers of moles of solvent, the two cratic contributions would differ. Thus $\mu_{HC}^\circ - \mu_W^\circ$ would not represent the true difference between the inherent properties of the solute in the two solvents. If mole fraction units are used, μ° represents only the internal free energy of the solute molecule, and the free energy of interaction with solvent and, since the internal free energy is the same in both solvents, $\mu_{HC}^\circ - \mu_W^\circ$ is exactly the quantity we want, the difference between the free energy of interactions with the solvent.

To relate μ_{HC}° to experimental data we use the analog of equation 1 for solution of a hydrocarbon in a hydrocarbon solvent

$$\mu_{HC} = \mu_{HC}^\circ + RT \ln X_{HC} + RT \ln f_{HC} \tag{2-2}$$

This equation can be applied to a dilute solution of one hydrocarbon in another. In principle a different value of μ_{HC}° would apply for each distinct solvent, but it is reasonable to assume that differences between similar solvents are small, that is, μ_{HC}° in, say *n*-hexane and *n*-decane, can be assumed to differ little from each other in comparison with the difference between μ_{HC}° and μ_W°. This means that equation 2-2 can be applied to a solution of a hydrocarbon in itself, that is, to the pure liquid hydrocarbon. In this case both $\ln X_{HC}$ and $\ln f$ would be zero and μ_{HC} would be equal to μ_{HC}°. This assumption of equivalence between different possible μ_{HC}° values is of course not rigorous, and small differences would undoubtedly be observed between different reference solvents if sufficiently accurate measurements were made.

ALIPHATIC HYDROCARBONS

Equations 2-1 and 2-2 may be applied to the extensive solubility measurements of liquid hydrocarbons in water at 25° by McAuliffe (1966). Since

$\mu_W = \mu^\circ_{HC}$ at saturation, $\mu^\circ_{HC} - \mu^\circ_W = RT \ln X_W$. Results for a homologous series of *n*-alkanes are shown in Fig. 2-1 and are seen to be a linear function of the length of the hydrocarbon chain. A least-squares treatment leads to the relation

$$\mu^\circ_{HC} - \mu^\circ_W = -2436 - 884 n_C \tag{2-3}$$

where n_C is the number of carbon atoms in the hydrocarbon chain and μ° is expressed in calories/mole. The variance in the constant term is ± 55 cal/mole, that in the slope is $\pm$ 13 cal/mole. The equation may be rewritten in terms of the contributions of the two terminal methyl groups and the internal CH_2 groups

$$\mu^\circ_{HC} - \mu^\circ_W = -2102 n_{CH_3} - 884 n_{CH_2} \tag{2-4}$$

but this must be done with the reservation that the equation applies only to straight-chain hydrocarbons, and that the coefficients of n_{CH_3} and n_{CH_2} do

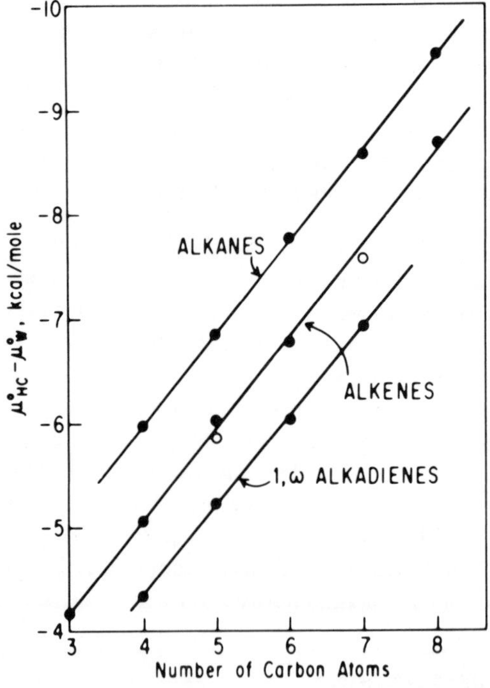

Fig. 2-1. Standard free energies for transfer of hydrocarbons from aqueous solution to pure liquid hydrocarbon at 25°*C*, based on solubility measurements of McAuliffe (1966). For the alkenes, filled circles represent 1-alkenes, open circles 2-alkenes.

not represent intrinsic measures of the hydrophobic effect for CH_3 and CH_2 groups on more complex hydrocarbons. Thus $\overset{\circ}{\mu}_{HC} - \overset{\circ}{\mu}_W$ for isobutane is -5850 cal/mole: if equation 2-4 were generally applicable, the three CH_3 groups alone would contribute -6300 cal/mole. Another example is provided by cyclopentane, for which $\overset{\circ}{\mu}_{HC} - \overset{\circ}{\mu}_W$ is -6000 cal/mole, compared to -4420 cal/mole as would be calculated from equation 2-4.

In fact, $\overset{\circ}{\mu}_{HC} - \overset{\circ}{\mu}_W$ is probably closely proportional to the number of water molecules that must be in contact with the dissolved hydrocarbon molecule in water solution. When straight-chain hydrocarbons are compared, the first member of the series is ethane, which would clearly have to make a much larger number of such contacts than the number of new contacts created by subsequent insertion of CH_2 groups between the methyl groups. On the other hand, replacement of a hydrogen atom by a methyl group, as in the formation of isobutane from propane, obviously creates only a small increment in solvent contacts (much of the new area of contact would involve H_2O molecules already adjacent to the terminal CH_3 groups of propane).

A regular relation for $\overset{\circ}{\mu}_{HC} - \overset{\circ}{\mu}_W$, such as equation 2-3 or 2-4, can thus be obtained only for homologous series in which addition of each carbon atom involves a regular increment in the extent of contact with solvent. It does apply, for example, to the series isobutane, 2-methyl butane, and 2-methyl pentane, with an average increment per added CH_2 group of -855 cal/mole. It also applies to unsaturated hydrocarbons. Figure 2-1 shows free energies calculated from solubilities of liquid *n*-alkenes and *n*-dienes in water (McAuliffe, 1966). It is seen that the position of the double bond in the chain has a small effect on the results, points for 2-alkenes lying slightly below those for 1-alkenes. A least-squares treatment of the combined data leads to the line drawn in the figure, corresponding to the relation

$$\overset{\circ}{\mu}_{HC} - \overset{\circ}{\mu}_W = -1503 - 884 n_C \qquad (2\text{-}5)$$

with a variance of $\pm$ 120 in the constant term and ± 22 in the slope. All the dienes investigated had double bonds in the terminal positions. The least-squares relation is

$$\overset{\circ}{\mu}_{HC} - \overset{\circ}{\mu}_W = -903 - 860 n_C \qquad (2\text{-}6)$$

with probable error about the same as for equation 2-3. It is obvious that introduction of one or more double bonds has considerable influence on the absolute value of $\overset{\circ}{\mu}_{HC} - \overset{\circ}{\mu}_W$, but virtually none on the dependence on n_C. In fact the constancy of the increment in $\overset{\circ}{\mu}_{HC} - \overset{\circ}{\mu}_W$ per added CH_2 group, as the chain length increases, or as branching points or double bonds are inserted, is quite remarkable. It must signify that interactions with solvents are of short range, and that interaction with water in particular cannot depend to a significant extent on the formation of cooperative linkages of water molecules with special space requirements.

The effect of unsaturation on the constant terms of equations 2-5 and 2-6 (compared to equation 2-3) indicates that an unsaturated hydrocarbon is less hydrophobic than a saturated one. Introduction of one double bond is essentially equivalent to removing one CH_2 group from a fully saturated chain. Introduction of a second double bond has a somewhat smaller effect, but all the data refer to dienes with terminal double bonds, and it is possible that larger effects would be observed (as indicated by the results for alkenes) if the double bonds were internal.

In equations 2-5 and 2-6 the organic solvent to which $\mathring{\mu}_{HC}$ refers (i.e., the pure liquid) is of course not a saturated hydrocarbon, and the results may therefore differ from those that would apply to solution in a saturated hydrocarbon. The increments per carbon atom indicate that any such difference is negligible insofar as CH_2 groups are concerned. The constant term might be affected more, as a result of possible interaction between one double bond and another. The available evidence indicates that any any effect from this source will be small. For example, μ° values for ethylene in *n*-hexane and in benzene differ by only 100 cal/mole.

AROMATIC HYDROCARBONS

Since we are primarily interested in these data in relation to biological membranes, the aromatic hydrocarbons are of peripheral importance only. Some results are tabulated in Table 2-1, and they show that aromatic hydrocarbons are considerably less hydrophobic than cyclic aliphatic hydrocarbons with the same number of carbon atoms. It must again be borne in mind that $\mathring{\mu}_{HC}$ refers to the pure liquid, and that $\mathring{\mu}_{HC} - \mathring{\mu}_{W}$ for transfer of an aromatic solute from water to a liquid saturated aliphatic hydrocarbon may differ from the figures given in this table. It is inconceivable however that this difference could amount to 2 kcal/mole, and the major factor responsible for the difference between cyclohexane and benzene must arise from interactions with water, that is, the hydrophobicity is less for benzene because its π-electrons lead to stronger van der Waals attraction to water molecules.

The effect of adding a methyl group to an aromatic ring is only slightly different from the effect of adding a methyl group to a cyclic aliphatic hydrocarbon; the difference between toluene and benzene is 810 cal/mole, the difference between methyl cyclohexane and cyclohexane is 900 cal/mole. The effect of adding a methyl group to a normal alkane depends on how close it is to an already existing methyl group (see earlier discussion): the difference between 2-methyl butane and butane is 750 cal/mole, that

Table 2-1. $\overset{\circ}{\mu}_{HC} - \overset{\circ}{\mu}_{W}$ for Aromatic and Saturated Cyclic Hydrocarbons at 25°C[a]

	$\overset{\circ}{\mu}_{HC} - \overset{\circ}{\mu}_{W}$ (cal/mole)
Cyclopentane	−6000
Methyl cyclopentane	−6880
Cyclohexane	−6730
Methyl cyclohexane	−7630
Benzene	−4620
Toluene	−5430
Ethylbenzene	−6230
o-xylene	−6180
m-xylene	−6110
p-xylene	−6100

[a] Based on solubilities of the liquid hydrocarbons in water as determined by McAuliffe (1966), except the results for *m*- and *p*-xylene, which are based on measurements of Bohon and Claussen (1951).

between 2-methyl pentane and pentane is 710 cal/mole, but the difference between 2,4-dimethyl pentane and 2-methyl pentane is 880 cal/mole.

SOLUBILITY IN AQUEOUS SALT SOLUTIONS

It is well known that salts decrease the solubility of nonpolar substances in water, which means that μ° in aqueous salt solution is more positive than $\overset{\circ}{\mu}_{W}$. However, the effect is appreciable at high salt concentrations only. On the basis of results for ethane, propane and butane (Morrison and Billett, 1952) the difference between μ° in a 1 molal salt solution and $\overset{\circ}{\mu}_{W}$ at 25°C is approximately given by the relations

$$
\begin{aligned}
\Delta\mu^\circ &= -231 n_{CH_3} - 38 n_{CH_2} && \text{for NaCl} \\
\Delta\mu^\circ &= -181 n_{CH_3} - 31 n_{CH_2} && \text{for LiCl} \qquad (2\text{-}7) \\
\Delta\mu^\circ &= -149 n_{CH_3} + 4 n_{CH_2} && \text{for KI}
\end{aligned}
$$

Differences between different salts are related to charge type and ionic radius.

The effect of salts on solubility is roughly proportional to salt concentration, and equations 2-7 thus show that $\Delta\mu^\circ$ is no larger than the uncertainty in $\overset{\circ}{\mu}_{HC} - \overset{\circ}{\mu}_{W}$ as given by equations 2-3 to 2-6 when the concentration of a uniunivalent salt is about 0.3 *M* or less. This means that we can assume the hydrophobic effect of *dilute* salt solutions to be essentially the same as that of pure water, and we shall therefore not make a distinction between them, that is, μ_W° will be used to refer to pure water or a dilute salt solution.

SOLUBILITY IN ALCOHOLS AND OTHER ORGANIC SOLVENTS

Nonpolar organic solvents such as CCl_4 are very close to hydrocarbons in their solvent properties. For example, $\mu_{CCl_4}^\circ - \mu_W^\circ$ for ethane is the same, within experimental error, as $\mu_{HC}^\circ - \mu_W^\circ$. However, if an organic solvent contains polar molecules, it will have some structural organization, and may resist the intrusion of nonpolar solute molecules. No known solvent has this property to the same extent as water does, but glycerol and ethylene glycol, for example, repel nonpolar substances quite effectively (Sinanoglu and Abdulnur, 1965).

The aliphatic alcohols are relatively benign solvents for hydrocarbons, presumably because the local molecular organization segregates the hydrocarbon tails from the OH groups, so that no major reorganization is required to accommodate hydrocarbon molecules, at least at low concentrations. Nevertheless, some reorganization is required and μ_{ROH}° (subscript ROH referring to alcohol as the solvent medium) is distinctly more positive than μ_{HC}°. This can be seen from the results in Table 2-2, which show the differ-

Table 2-2. Free Energy of Transfer from Alcohols to Water at 25°C[a]

		$\mu_{ROH}^\circ - \mu_{HC}^\circ$	$\mu_{ROH}^\circ - \mu_W^\circ$	$\mu_{HC}^\circ - \mu_W^\circ$
Hydrocarbon	Solvent		cal/mole	
Propane	MeOH	1460	3620	5080
	EtOH	1100	3980	
	isoPrOH	880	4200	
Butane	MeOH	1600	4370	5970
	EtOH	1190	4780	
	isoPrOH	940	5030	
	Glycol[a]	3700	2300	
	Glycerol[a]	4800	1200	

[a] The figures for butane in glycol and glycerol are based on data of uncertain reliability, taken from the Landolt-Börnstein Tabellen (1960). All other results are based on the work of Kretschmer and Wiebe (1952), and represent accurate data, extrapolated to infinite dilution of the hydrocarbon in the alcohol solution.

ence between $\overset{\circ}{\mu}_{\text{ROH}}$ and $\overset{\circ}{\mu}_{HC}$ for propane and butane in several alcohols. It is seen that the difference is relatively small and decreases as the hydrocarbon portion of the alcohol increases in size. In the last two columns of this table values of $\overset{\circ}{\mu}_{\text{ROH}} - \overset{\circ}{\mu}_{W}$ are compared with values of $\overset{\circ}{\mu}_{HC} - \overset{\circ}{\mu}_{W}$, and it is seen, with reference to equation 2-4 or 2-5, that the difference between these quantities lies more in the constant term of the equation than in the increment per carbon atom. The difference between $\overset{\circ}{\mu}_{\text{ROH}} - \overset{\circ}{\mu}_{W}$ for butane and propane is 750, 800, and 830 cal/mole, respectively, in the three alcohols. The corresponding difference in $\overset{\circ}{\mu}_{HC} - \overset{\circ}{\mu}_{W}$ is 890 cal/mole.

Table 2-2 contains some rough data for glycerol and ethylene glycol, which show that a strong solvophobic effect is associated with these solvents, especially with glycerol, which is almost as antagonistic to butane as water is.

REFERENCES

Bohon, R. L., and W. F. Claussen. (1951). *J. Am. Chem. Soc.*, **73**, 1571.

Gurney, R. W. (1953). *Ionic Processes in Solution.* McGraw-Hill, New York. (Reprinted by Dover Publications, New York, 1962.)

Kąuzmann, W. (1959). *Adv. Protein Chem.*, **14**, 1.

Kretschmer, C. B., and R. Wiebe. (1952). *J. Am. Chem. Soc.*, **74**, 1276.

Landolt-Börnstein Tabellen (1962). 6th ed., vol. II, section 2b, Springer-Verlag, Berlin.

McAuliffe, C. (1966). *J. Phys. Chem.*, **70**, 1267.

Morrison, T. J., and F. Billett. (1952). *J. Chem. Soc.*, 3819.

Sinanoglu, O., and S. Abdulnur. (1965). *Fed. Proc.*, **24**, S-12.

THREE

SOLUBILITY OF AMPHIPHILES IN WATER AND ORGANIC SOLVENTS

Ionic or polar derivatives of hydrocarbons were called *amphipathic* by Hartley (1936) because they contain one part that has "sympathy" and another that has "antipathy" for water. We have preferred in this book to use the alternative and more euphonious term *amphiphilic.* To avoid repetition of the clumsy phrase "amphiphilic molecule or ion," we shall use the noun *amphiphile* as designating either an ionic or a molecular species. The polar or ionic group of an amphiphile is generally called the *head*, and the hydrocarbon chain is called the *tail.* Because solvent-solute interactions are of quite short range, the solvent-dependent part of the chemical potential of such molecules may be thought of as the sum of nearly independent contributions from the tail and the head. As the length of the tail is increased, its contribution becomes dominant, and the amphiphile becomes progressively less soluble in aqueous solutions.

At one position on an amphiphilic ion or molecule, the point of attachment of the tail to the head, the hydrophilic and hydrophobic properties are in conflict. Since the ordering of water molecules by the polar head is the result of strong attractive forces, the influence of the hydrophilic head group is likely to predominate. Thus one CH_2 group of the hydrophobic tail (proximal to the head group) may be expected to make little or no contribution to the hydrophobic effect, and one or two additional CH_2 groups may conceivably be affected to a lesser degree. It does not seem probable that the influence of the head group can make itself felt at any greater distance from the point of attachment.

The equilibrium distribution of fatty acids between *n*-heptane and aqueous buffer solutions, ionic strength 0.1 to 0.15, has been studied by Goodman (1958) and by Smith and Tanford (1973). As noted earlier, the presence of salts at relatively low concentrations should not significantly affect hydrocarbon-water interactions and the results can be considered equivalent to those that would have been obtained if the aqueous phase had been pure water. Using equation 2-1 for the chemical potential of the solute in the aqueous solution and equation 2-2 for the solution in heptane, we have at equilibrium

$$\mathring{\mu}_{HC} - \mathring{\mu}_{W} = RT \ln X_W/X_{HC} + RT \ln f_W/f_{HC} \qquad (3\text{-}1)$$

and the desired free energy difference can be obtained from the distribution ratio X_W/X_{HC}.

There are experimental complications, in that X_W/X_{HC} is very small for acids with long alkyl chains. To have a measurable concentration of solute in the aqueous phase while keeping X_{HC} reasonably small, it is necessary to use a high pH, so that most of the solute in the aqueous phase will be in the form of the anion, $RCOO^-$, the actual value of X_W being calculated from the measured concentration by use of the acid dissociation constant. Furthermore, there is a strong tendency for association in both solvents: in the organic phase, dimers of RCOOH are formed with hydrogen bonds between the head groups, and in the aqueous phase there is strong association between RCOOH and $RCOO^-$. As a result, considerable care has to be exercised to ascertain that experimental data are extended to conditions where the activity coefficient term in equation 3-1 can be neglected.

The results obtained are shown in Fig. 3-1. The earlier data of Goodman (1958) appeared to indicate that a linear relation between $\mathring{\mu}_{HC} - \mathring{\mu}_{W}$ and alkyl chain length extends only as far as palmitic acid, and this led Mukerjee (1967) to suggest that there might be a limit to the hydrophobicity of a hydrocarbon chain, possibly indicating that chains with more than 16 carbon atoms fold upon themselves in aqueous solution, so that no appreciable further increase in the hydrocarbon-water interface occurs. The recent data of Smith and Tanford (1973) have shown, however, that Goodman's results for palmitic and stearic acids were obtained under conditions where association between RCOOH and $RCOO^-$ in the aqueous phase still affects the distribution. When this difficulty is avoided the linear relation between $\mathring{\mu}_{HC} - \mathring{\mu}_{W}$ persists at least up to behenic acid (22 carbon atoms). Least-squares analysis of the data, ignoring Goodman's results for the higher acids, leads to the relation (in cal/mole)

$$\mathring{\mu}_{HC} - \mathring{\mu}_{W} = 4260 - 825 n_C \qquad (3\text{-}2)$$

The constant term on the right-hand side of equation 3-2 is positive, reflecting the high polarity and consequent preference for water of the COOH

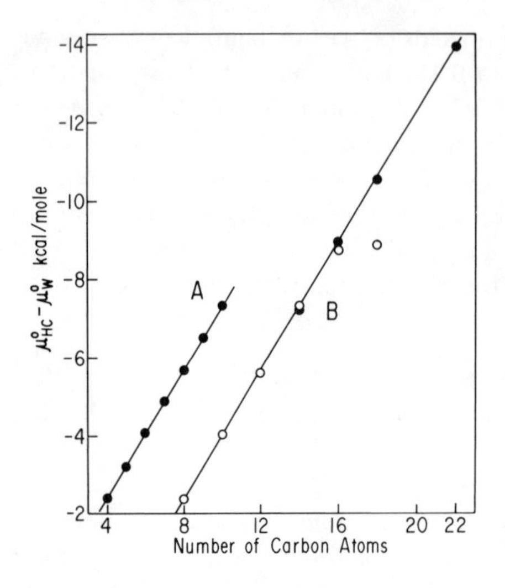

Fig. 3-1. Curve A represents free energies for transfer of pure aliphatic alcohols from aqueous solution to pure liquid alcohol at 25°C, based on data of Kinoshita et al. (1958). Curve B represents free energies of transfer of undissociated fatty acids from a dilute aqueous buffer solution to liquid *n*-heptane at 23 to 25°C. Open circles represent data of Goodman (1958), filled circles data of Smith and Tanford (1973).

group. If we assume that the terminal methyl group of the alkyl chain makes the same contribution to $\mu^o_{HC} - \mu^o_W$ as in a pure hydrocarbon (− 2100 cal/mole) and that one CH_2 group contributes nothing because the water around it is associated (in the aqueous phase) with the head group, the actual contribution of the COOH group to $\mu^o_{HC} - \mu^o_W$ would be +3900 cal/mole.

The dependence of $\mu^o_{HC} - \mu^o_W$ on n_C reflects the hydrophobic interactions of CH_2 groups inserted in the chain as the chain length is increased from 8 to 22 carbon atoms. The value obtained, 825 cal/mole, is somewhat smaller than the corresponding figures from equations 2-3, 2-5, and 2-6, but the difference is not much greater than the experimental uncertainty.

In addition to the results shown in Fig. 3-1, Goodman (1958) obtained distribution data for oleic and linoleic acids, which are C_{18} acids with one and two double bonds, respectively. The results indicate that the introduction of double bonds decreases the magnitude of hydrophobic interactions, as was observed with pure hydrocarbons. A quantitative comparison cannot be made because of the previously mentioned error in Goodman's results for stearic acid.

Figure 3-1 also shows data based on the solubilities of aliphatic alcohols in water (Kinoshita et al., 1958). The saturating phase was the pure liquid alcohol, and the chemical potential in that phase is represented by μ^o_{ROH}. The results may be expressed as

$$\mu^o_{ROH} - \mu^o_W = 833 - 821n_C \tag{3-3}$$

with a variance of only ± 3 cal/mole in the slope, and ±25 cal/mole in the constant term. It is seen that the increment per CH_2 group is the same as that in equation 3-2, even though the organic solvent in this case is not a pure hydrocarbon. This agreement may be attributed to the probable hydrogen bonding between OH groups in a liquid alcohol, which would leave the distal portions of hydrocarbon tails essentially in a hydrocarbon environment. (See discussion of solubilities of hydrocarbons in alcohols in Chapter 2.) The constant term in equation 3-3 is very much smaller than that for carboxylic acids, given in equation 3-2, indicating that the terminal OH group is less hydrophilic than a COOH group, which is not surprising, since the latter can form a larger number of hydrogen bonds. Taking the contribution of the terminal methyl group to $\mu^{\circ}_{ROH} - \mu^{\circ}_{W}$ to be -2100 cal/mole, as before, we find that the contribution of the terminal OH group becomes $+1340$ cal/mole if the CH_2 group proximal to the OH group is assumed to make no contribution to $\mu^{\circ}_{ROH} - \mu^{\circ}_{W}$.

These results support the idea expressed at the beginning of this chapter, that the contributions of the hydrophilic and hydrophobic portions of an amphiphile to the free energy of interaction with solvent should be nearly independent. One might have expected the increment per carbon atom in equation 3-2 to be a few percent larger, that is, the same as in equations 2-3, 2-5, and 2-6. However, it must be recognized that we are dealing with isolated experimental studies and that the absolute accuracy of the measurements may not be as good as the self-consistency expressed by the close adherence to the linear relations. A somewhat smaller increment per carbon atom in equation 3-3 is theoretically not unexpected because the organic solvents in this case are a series of alcohols.

Reference should be made to a study by Schrier and Schrier (1967), which indicates that the effect of added salt on μ°_{W} of organic amides is also an additive function of approximately independent contributions from polar and nonpolar groups.

REFERENCES

Goodman, D. S. (1958). *J. Am. Chem. Soc.*, **80**, 3887.
Hartley, G. S. (1936). *Aqueous Solutions of Paraffin-Chain Salts*. Hermann & Cie., Paris.
Kinoshita, K., H. Ishikawa, and K. Shinoda. (1958). *Bull. Chem. Soc. Japan*, **31**, 1081.
Mukerjee, P. (1967). *Adv. in Colloid and Interface Sci.*, **1**, 241.
Schrier, E. E., and E. B. Schrier. (1967). *J. Phys. Chem.*, **71**, 1851.
Smith, R., and C. Tanford. (1973). *Proc. Nat. Acad. Sci. U.S.A.*, **70**, 289.

FOUR

THE EFFECT OF TEMPERATURE

ANOMALOUS ENTROPY AND HEAT CAPACITY

The effect of temperature on the free energy of any process allows one to determine the separate contributions of energy (or enthalpy) and of entropy to the free energy. This is often very helpful in arriving at a qualitative molecular mechanism for the process, and this is especially true for the hydrophobic effect. The classical mechanistic representation of the hydrophobic effect by Frank and Evans (1945) is based primarily on such considerations, and more sophisticated approaches to the problem have been able to add little to the original picture, for reasons to be discussed in Chapter 5. In this chapter we summarize the essential points made by Frank and Evans (insofar as they refer to hydrocarbon chains), basing the discussion, however, on more recent experimental data.

The relation between $\overset{\circ}{\mu}_{HC} - \overset{\circ}{\mu}_W$ and the corresponding partial molal enthalpies and entropies is

$$\overset{\circ}{\mu}_{HC} - \overset{\circ}{\mu}_W = \bar{H}^\circ_{HC} - \bar{H}^\circ_W - T(\bar{S}^\circ_{HC} - \bar{S}^\circ_W) \tag{4-1}$$

The separate contributions can be evaluated if the free energy difference is determined as a function of temperature, since

$$\frac{d[(\overset{\circ}{\mu}_{HC} - \overset{\circ}{\mu}_W)/T]}{d(1/T)} = \bar{H}^\circ_{HC} - \bar{H}^\circ_W \tag{4-2}$$

Alternatively, the enthalpy difference can be measured directly by calorimetry. In either case, the entropy difference is obtained by difference from equation 4-1. Since we shall use the data as a basis for discussing the state of the hydrocarbon in the aqueous medium, it will be convenient to consider the thermodynamic functions as they apply to the transfer process in the direction opposite to that considered thus far, that is, as $\mu_W^\circ - \mu_{HC}^\circ$, $\overline{H}_W^\circ - \overline{H}_{HC}^\circ$, and so on.

The hydrocarbon solubilities of Fig. 2-1 have not been determined as functions of temperature, and the corresponding enthalpies and entropies can therefore not be evaluated, but data exist for some comparable systems. The most important features are illustrated by Fig. 4-1, which is based on relative solubilities of gaseous ethane in water and in two organic solvents, benzene and carbon tetrachloride. As was indicated in Chapter 2, μ° values in such solvents (which we shall designate as μ_{org}°) differ only minimally from μ_{HC}°. For ethane, $\mu_W^\circ - \mu_{org}^\circ$ is + 3800 and +3600 cal/mole in CCl_4 and in benzene, respectively. The value of $\mu_W^\circ - \mu_{HC}^\circ$ is not known with precision: its probable value is +3900 cal/mole.

The results obtained from Fig. 4-1 are tabulated in Table 4-1, and they are quite unexpected in two ways. (1) The solubility in water decreases with increasing temperature, and $\overline{H}_W^\circ - \overline{H}_{org}^\circ$ is therefore negative. This means

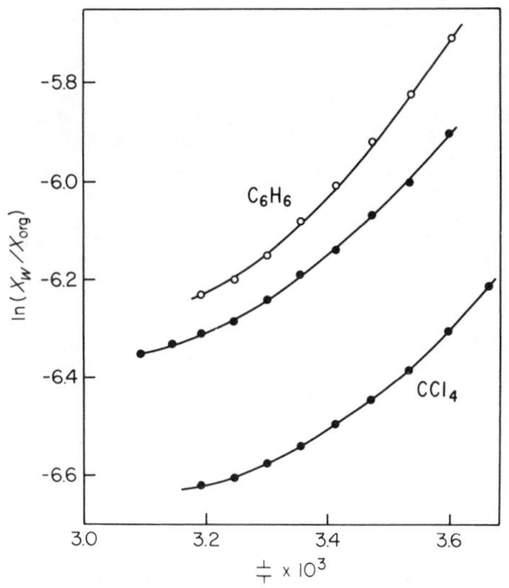

Fig. 4-1. Van't Hoff plot for the relative solubilities of ethane in water and in two organic solvents. Solubilities in the organic solvents are based on the work of Horiuti, as given in the Landolt-Börnstein Tabellen (1962). Solubilities in water are taken from Morrison (1952). For benzene, an alternative calculation, based on water solubilities by Claussen and Polglase (1952) is shown by means of open circles.

Table 4-1. Thermodynamic Parameters for Transfer of Hydrocarbons from Organic Solvents to Water

	From CCl_4 or C_6H_6 to Water				
	$\mu^\circ_W - \mu^\circ_{org}$ (cal/mole)	$\bar{H}^\circ_W - \bar{H}^\circ_{org}$ (cal/mole)	$\bar{S}^\circ_W - \bar{S}^\circ_{org}$ (cal/deg mole)	$(\bar{C}^\circ_p)_W - (\bar{C}^\circ_p)_{org}$ (cal/deg mole)	$(\bar{C}^\circ_p)_{gas}$ (cal/deg mole)
C_2H_6[a]	3800	−1800	−19	59	12.6
C_2H_6[b]	3600	−2200	−20	59	12.6

	From Pure Liquid Hydrocarbon to Water[c]				
	$\mu^\circ_W - \mu^\circ_{HC}$ (cal/mole)	$\bar{H}^\circ_W - \bar{H}^\circ_{HC}$ (cal/mole)	$\bar{S}^\circ_W - \bar{S}^\circ_{HC}$ (cal/deg mole)	$(\bar{C}^\circ_p)_W - (\bar{C}^\circ_p)_{HC}$ (cal/deg mole)	$(\bar{C}^\circ_p)_{HC}$ (cal/deg mole)
C_2H_6	3900	−2500	−21	—	—
C_3H_8	4900	−1700	−22	—	—
C_4H_{10}	5900	− 800	−23	65	35
C_6H_6	4600	+ 600	−13	108	32
$C_6H_5CH_3$	5300	+ 600	−16	108	37
$C_6H_5C_2H_5$	6100	+ 400	−19	108	43

[a] From CCl_4, data from Fig. 4-1.
[b] From C_6H_6, data from Fig. 4-1.
[c] Results for the three aliphatic hydrocarbons are based on solubilities of the gaseous hydrocarbons in water (Morrison, 1952; Claussen and Polglase, 1952; Kresheck et al., 1965) together with extrapolated parameters for the vaporization of the corresponding liquids (American Petroleum Institute, 1953). Data for the aromatic hydrocarbons are based on their solubilities in water as a function of temperature (Bohon and Claussen, 1951).

that the transfer of the hydrocarbon to water is *energetically favored*, and that the positive value of $\mu^\circ_W - \mu^\circ_{org}$ results from a large negative entropy change in the process. Since the data are in unitary units, this must be an inherent entropy change involving an ordering process in the solution, and cannot be associated with dispersal of the solute per se. (2) The van't Hoff plots are curved, to an extent that is very unusual for simple processes such as this. This means that the enthalpy change is itself strongly temperature dependent, and this in turn tells us that a large change in heat capacity accompanies the process, since

$$\frac{d(\bar{H}^\circ_W - \bar{H}^\circ_{org})}{dT} = (\bar{C}^\circ_P)_W - (\bar{C}^\circ_P)_{org} \tag{4-3}$$

Table 4-1 contains, in addition to the results derived from Fig. 4-1, data for the transfer of three aliphatic hydrocarbons from the pure liquid state to water. These results are intrinsically less reliable than those based on the figure: they are obtained by combining solubilities of the gaseous hydrocarbons in water with extrapolated parameters for the vaporization of the corresponding liquids. For ethane and propane a long extrapolation is necessary. (Note, however, that even the more direct data of Fig. 4-1 are not very precise, as indicated by the appreciable difference between measurements of water solubility as determined in two different laboratories.) Table 4-1 also contains comparable data for three aromatic hydrocarbons, based on solubilities of the liquid hydrocarbons in water.

The enthalpy, entropy, and heat capacity values of amphiphilic molecules in water display similar features to those characteristic of pure hydrocarbons, as has been recognized for a long time. The anomalous heat capacities in particular are easily recognized from direct measurements, and attention to them was first drawn by Edsall (1935). Table 4-2 provides experimental data for some aliphatic alcohols. The enthalpy and heat capacity data were obtained by direct calorimetric measurement. For the saturated alcohols, the corresponding free energies of transfer are given by equation 3-3, and they have been used to obtain entropies of transfer by use of equation 4-1.

The negative enthalpies and entropies of Tables 4-1 and 4-2, and the very large anomalous heat capacities, provide considerable insight into the origin of the hydrophobic effect, which will be briefly discussed. It is apparent that the heat capacities for transfer to water are much larger than the heat capacities of the solute molecules themselves, as seen from the values for gaseous ethane and for other hydrocarbons in the liquid state, listed in the last column of Table 4-1. There is no possibility that such large increases in $\bar{C}_p$ can result from the effect of temperature on the internal motions of the solute molecules. They must be associated with changes in the state of water molecules brought about by the presence of the dissolved hydrocarbon.

INTERPRETATION IN TERMS OF CHANGES IN WATER STRUCTURE

Water is a highly structured liquid, with hydrogen bonds linking the individual molecules to each other. The precise arrangement about each molecule is not known (see Chapter 5), but the tetrahedral symmetry of the oxygen bond orbitals and the tetrahedral structure of ice suggest a locally tetrahedral arrangement of molecules in the liquid state also. Any arrangement of this type would have to be disrupted by *any solute* dissolved in water. Some hydrogen bonds would have to be broken: if the solute is polar, new

Table 4-2. Thermodynamic Parameters for Transfer of Aliphatic Alcohols from the Pure Liquid to Water at 25°C[a]

	$\mu_W^\circ - \mu_{ROH}^\circ$ (cal/mole)	$\bar{H}_W^\circ - \bar{H}_{ROH}^\circ$ (cal/mole)	$\bar{S}_W^\circ - \bar{S}_{ROH}^\circ$ (cal/deg mole)	$(\bar{C}_p^\circ)_W - (\bar{C}_p^\circ)_{ROH}$ (cal/deg mole)
		Saturated Alcohols		
Ethanol	760	−2430	−10.7	39
n-propanol	1580	−2420	−13.4	56
n-butanol	2400	−2250	−15.6	72
n-pentanol	3222	−1870	−17.1	84
		Unsaturated Alcohols		
2-propen-1-ol		−1700		47
2-buten-1-ol		−1940		57
3-buten-1-ol		−2000		57
4-penten-1-ol		−1410		58

[a] Enthalpies and heat capacities were determined calorimetrically by Arnett et al. (1969). Values of $\mu_W^\circ - \mu_{ROH}^\circ$ for the saturated alcohols were calculated by equation 3-3, and it should be noted that the direct data on which that equation is based extend only from butanol to decanol.

hydrogen bonds between water and solute would be formed, but if the solute is nonpolar, such as a hydrocarbon, and unable to form hydrogen bonds, the net result expected is an increase in the enthalpy of the solution and, thereby, a positive value for $\mu_W^\circ - \mu_{org}^\circ$. The intrinsic difference between water and alcohols in this regard is worth noting. In a liquid alcohol the tendency for O—H···O hydrogen bond formation is likely to be as strong as in water, but such hydrogen bonds cannot comprise an isotropic network filling the entire solvent space. Hydrogen-bonded regions must alternate with regions composed of alkyl side chains, and a nonpolar solute can be dissolved in such a region without major disruption of solvent organization and, certainly, without disruption or appreciable distortion of hydrogen bonds.

It is evident from the results in Tables 4-1 and 4-2, and similar data for other nonpolar and amphiphilic molecules, that the intuitive explanation for the hydrophobic effect given in the previous paragraph is not adequate. The rupture of hydrogen bonds requires energy and, if the preceding explanation were correct, the major contribution to the free energy of transfer to water would be a positive enthalpy term. Instead, as we have seen, transfer to water is actually slightly favored energetically, the positive value of

$\overset{\circ}{\mu}_W - \overset{\circ}{\mu}_{org}$ resulting from a negative entropy change for the process. There is only one possible explanation, as was pointed out in a classic paper by Frank and Evans (1945): the water molecules at the surface of the cavity created by a nonpolar solute must be capable of rearranging themselves in order to regenerate the broken hydrogen bonds (in fact, the negative enthalpy change indicates they are slightly stronger than before), but in doing so they create a higher degree of local order than exists in pure liquid water, thereby producing a decrease in entropy.

The results show, furthermore, that this local ordering of water molecules, for solutes containing hydrocarbon chains, is not a regular function of hydrocarbon chain length. Whereas $\overset{\circ}{\mu}_W - \overset{\circ}{\mu}_{HC}$ or $\overset{\circ}{\mu}_W - \overset{\circ}{\mu}_{org}$ increases linearly with increasing hydrocarbon chain length, and seems to depend primarily on the magnitude of the hydrocarbon-water interfacial area, the corresponding enthalpy and entropy functions show no corresponding regularity. This indicates that the water molecules at a hydrocarbon-water interface do not have a unique way of arranging themselves, but that different arrangements are possible, depending on the precise spatial requirements. These different arrangements must differ in enthalpy and entropy, but must do so in a mutually compensating fashion, so that no irregularity in free energy can be detected.

The anomalous heat capacities are consistent with the existence of such multiple arrangements. If two such arrangements have nearly the same free energy, they must be about equally probable, that is, there would be a local equilibrium between two states of the solvent, one with a higher enthalpy *and* a higher entropy than the other. An increase in temperature would shift the equilibrium toward this higher enthalpy state, and lead to a higher enthalpy for the solution as a whole. Experimentally, this enthalpy increase would manifest itself as an anomalous contribution to the partial molal heat capacity $\bar{C}_p^{\circ}$ of the solute in the aqueous solution. One possibility is that in the "high enthalpy" state the cavity containing the solute is lined by broken hydrogen bonds (as in the intuitive model that these data have forced us to discard), but it is equally likely that the hydrogen bonds are weaker and less rigidly ordered, without any change in their number. The main conclusion is that the organization of water at a hydrocarbon-water interface is a makeshift arrangement, subject to alteration as a result of temperature changes as well as changes in hydrocarbon chain length.

SOLUTIONS OF IONS IN WATER

As was noted above, *any solute* must disrupt the local ordering of solvent molecules in liquid water. It is therefore instructive, as a further aid to visualizing the structural implications of thermodynamic data for solutions of hydrocarbons in water, to compare these data with those for other solutes. Solutions of ions are particularly interesting in this regard, because they are also frequently characterized by negative entropy contributions, indicative of the ordering of water molecules around the ions by ion-dipole forces. An example is provided by the reaction $H_2O + CH_3COOH \rightarrow H_3O^+ + CH_3COO^-$. The enthalpy change in this reaction is close to zero, because there is essentially no net change in the number or strength of hydrogen bonds. The value of $\Delta S°$ for the reaction at 25°C (in unitary units) is − 30 cal/mole deg and, as before, the only conceivable origin for this lies in the organization of solvent molecules. In this case, however, ΔC_p° for the reaction is also negative (−34 cal/deg-mole), pointing to the intrinsic difference between the modes of organization induced by nonpolar and polar solutes. The water molecules near a nonpolar solute are hydrogen-bonded to each other in an intrinsically labile state (as indicated above), *more* labile than the ordered arrangement of water molecules in pure water. The water molecules surrounding an ion are firmly hydrogen-bonded, and *less* labile than they would be in pure water.

The picture is somewhat complicated by the fact that some metal ions have an even more negative partial molal heat capacity than can be accounted for in this way, but such ions also have an anomalous *positive* component to the partial molal entropy. Frank and Evans (1945) have interpreted this phenomenon as indicating the existence of a layer of disordered water molecules at the interface between a hydrated ion and the surrounding solvent.

THE SURFACE FREE ENERGY OF WATER

Water in contact with air tends to minimize its surface area by forming drops, that is, the transfer of a water molecule from bulk water to the surface is accompanied by an increase in free energy. Moreover, the surface formed has considerable two-dimensional strength: hydrogen bonds parallel to the surface undoubtedly form a stronger network than exist in any plane in bulk water. Intuitively this would seem to be a process resembling that which occurs at the interface between water and a hydrophobic solute. However, caution must be exercised in drawing a parallel between the two processes, for the enthalpy and entropy changes accompanying formation of a water-air interface are both positive, that is, although the hydrogen bonds in the plane of the surface of a drop of water are very strong, the total energy of a water molecule in the surface is greater, not less than the energy per molecule within the liquid.

REFERENCES

American Petroleum Institute (1953). *Selected Values of Physical and Thermodynamic Properties of Hydrocarbons and Related Compounds.* Carnegie Press.
Arnett, E. M., W. B. Kover, and J. V. Carter. (1969). *J. Am. Chem. Soc.*, **91**, 4028.
Bohon, R. L., and W. F. Claussen. (1951). *J. Am. Chem. Soc.*, **73**, 1571.
Claussen, W. F., and M. F. Polglase. (1952). *J. Am. Chem. Soc.*, **74**, 4817.
Edsall, J. T. (1935). *J. Am. Chem. Soc.*, **57**, 1506.
Frank, H. S., and M. W. Evans. (1945). *J. Chem. Phys.*, **13**, 507.
Kresheck, G. C., H. Schneider, and H. A. Scheraga. (1965). *J. Phys. Chem.*, **69**, 3132.
Landolt-Börnstein Tabellen (1963). 6th ed., vol. II, secton 2b, Springer-Verlag, Berlin.
Morrison, T. J. (1952). *J. Chem. Soc.*, 3814.

FIVE

THE STRUCTURE OF WATER[1]

We have seen that a qualitative picture of the structural changes that occur in water when a hydrophobic solute is added can be deduced from thermodynamic data alone. We should like of course to have a more definite, quantitative picture of the molecular rearrangements that take place, but to do so we must first have definite knowledge of the structure of pure liquid water. This chapter will present a very brief summary of present knowledge of the structure of water, and will show that our information is not complete enough to warrant an attempt to formulate a precise structural model for aqueous solutions.

ORDINARY ICE AND OTHER CRYSTALLINE FORMS

The most powerful tools for elucidation of exact structural information are X-ray and neutron diffraction of crystalline substances, and these methods have of course been applied to determine the structures of ordinary ice and of the other, denser crystal forms of water that are stabilized by high pressure. They have also been used to determine the structures of crystalline complexes of water with small nonpolar molecules (such as $Cl_2 \cdot 8\ H_2O$ and $CH_4 \cdot 5\frac{3}{4}\ H_2O$) that are formed when water is crystallized in the presence of these substances. The latter are of special interest to us since they allow us to see some possible arrangements of water molecules in the presence of hydrophobic molecules. The structures of ordinary ice and of one of the gas hydrates are shown in Figs. 5-1 and 5-2.

The hydrides of the smaller nonmetallic elements are gases at room tem-

[1] Most of this chapter is based on a recent book on the structure and properties of pure water by Eisenberg and Kauzmann (1969).

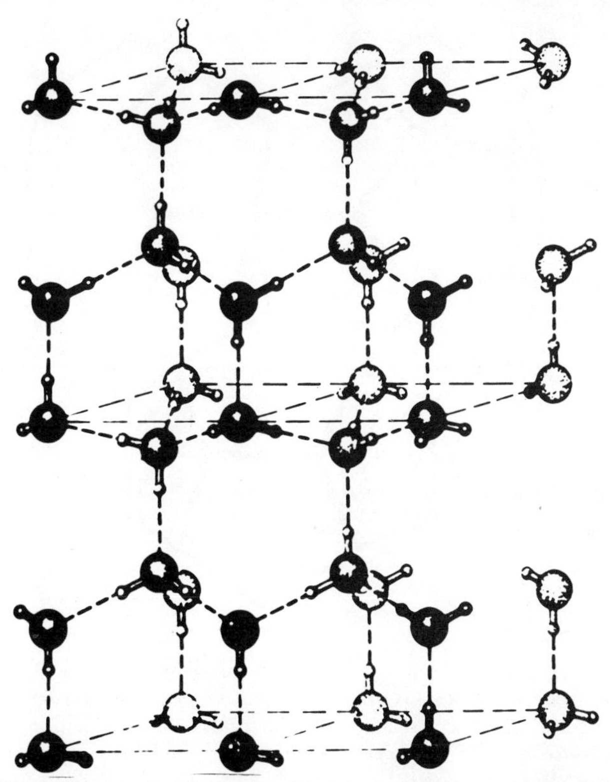

Fig. 5-1. The structure of ordinary ice. Large spheres represent oxygen atoms, small spheres hydrogen atoms. The orientation of water molecules is arbitrary: alternative arrangements can arise from rotation of water molecules. Reprinted from Linus Pauling: *The Nature of the Chemical Bond.* Copyright 1939 and 1940 by Cornell University. Third edition © 1960 by Cornell University. Used by permission of Cornell University Press.

perature. H_2O is the sole exception, and its existence in condensed phases is due to the strength of O—H···O hydrogen bonds, and to the fact that each water molecule can form four such bonds, two in which it acts as a hydrogen donor and two in which it acts as a receptor. As a consequence of this arrangement, the crystal structure adopted by ordinary ice (Fig. 5-1) is a tetrahedral one, with a distance of about 2.76 Å between neighboring oxygen atoms. This arrangement is a relatively open one. The nearest nonhydrogen-bonded neighbors are 4.5 Å apart, and much unfilled space is left within the crystal. This feature accounts for the low density of ice, one of relatively few crystalline materials that becomes more dense when it melts.[2] It should be noted

[2] This property is shared by other substances with a crystal structure based on tetrahedral contacts with only four nearest neighbors. Most of the examples (e.g., diamond) are substances with high melting points.

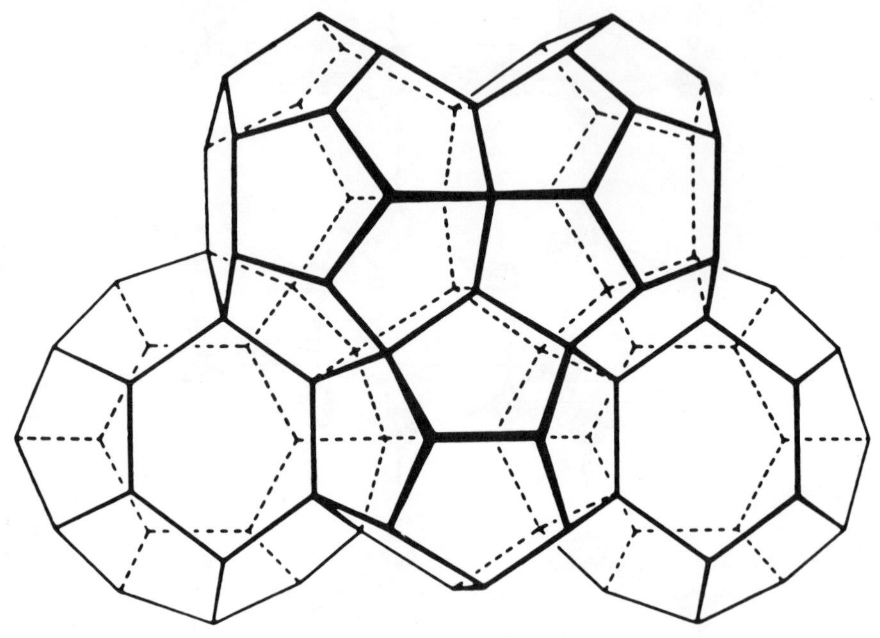

Fig. 5-2. A portion of the hydrogen bond framework of crystalline chlorine hydrate: each line represents an O—H···O bond. Molecules of Cl_2 are contained in the larger (tetrakaidecahedral) cavities, but not in the dodecahedral ones. Reprinted from Linus Pauling: *The Nature of the Chemical Bond.* Copyright 1939 and 1940 by Cornell University. Third edition © 1960 by Cornell University. Used by permission of Cornell University Press.

that the H_2O molecule remains intact in the crystal: the H atoms do not move to positions halfway between adjacent O atoms, but remain about 1.00 Å from the O atom to which they "belong" and 1.76 Å from the O atom to which they form a hydrogen bond. These specifications still permit a vast number of possible orientational arrangements (which may be altered by cooperative rotations of the water molecules), and the crystals have been found to be disordered with respect to this feature, as was first suggested by Pauling (1935) on the basis of the residual entropy of ice at 0°K.

The unfilled space within normal ice crystals allows ice to respond to increased pressure by formation of higher density structures. Eight stable polymorphic ordered forms are known, ranging in density up to 1.66 g/cc (Kamb, 1968). The water molecules in these denser crystals retain the basic hydrogen-bonded structure, with each water molecule hydrogen-bonded to its four nearest neighbors. However, some of the hydrogen bonds are distorted (O—H···O is no longer linear), and nonbonded water molecules are thereby permitted to approach much closer to each other than in ordinary ice: in ice II the closest nonhydrogen bonded neighbors are 3.24 Å apart, in the densest polymorphs (ice VII and ice VIII) the distance between nonbonded neighbors and hydrogen-bonded neighbors becomes equal at 2.86 Å.

Distortion of the tetrahedral structure also occurs in the crystalline hydrates of nonpolar solutes. In this case the distortion serves to make room for solute molecules instead of nonbonded water molecules. The structure of one such hydrate, $Cl_2 \cdot 8\ H_2O$, has been determined in detail by X-ray crystallography (Pauling and Marsh, 1952) and is shown in part in Fig. 5-2. The basic element of the structure is a pentagonal dodecahedron. The dodecahedra are arranged (together with six interstitial water molecules per four dodecahedra) so as to define tetrakaidecahedral as well as dodecahedral cavities. The O—H···O distance is 2.75 Å, and O—O—O bond angles range from 106 to 125°. A chlorine molecule is too large to fit into the dodecahedral spaces, but it is readily accommodated by the larger tetrakaidecahedral cavities. The calculated composition based on the assumption that only the latter contain Cl_2 is $Cl_2 \cdot 7\frac{2}{3}\ H_2O$, which is in good agreement with the analytical composition, $Cl_2 \cdot 8\ H_2O$. An interesting feature of the structure is that molecules somewhat smaller than Cl_2 could be accommodated by the dodecahedra as well, and the composition would then be $X \cdot 5\frac{3}{4}\ H_2O$. This is exactly the composition of crystalline xenon and CH_4 hydrates, for which detailed crystal structures are not available.

A similar structural reorganization is observed in the crystalline hydrates of some tetraalkyl ammonium salts. One of these, $(iso\text{-}C_5H_{11})_4N^+F^- \cdot 38\ H_2O$, has been investigated in detail by Feil and Jeffrey (1961). The organic cations are found to be contained in water "cages" formed by the faces of pentagonal dodecahedra. The distance between nearest neighbor oxygen atoms ranges from 2.75 to 2.86 Å in different parts of the structure, and the O—O—O bond angle ranges from 91 to 133°.

LIQUID WATER

We are of course interested in the liquid, not the crystalline state of water. Direct structural information comparable to that for the crystalline state is not available for the liquid state, or it might be better to say that a structure in that sense does not exist in the liquid state, in which there can be order at very short range only. This short-range order, however, might be expected to resemble the short-range order in crystalline states since, as was indicated above, the latter is largely dictated by the properties of the H_2O molecule per se. Indeed, all information that bears upon the structure in the liquid state confirms this expectation, and the actual problem in discussing liquid water is to decide between alternative ways in which the local ordering of water molecules in ice might be modified in the melting process. We shall briefly discuss some of the experimental data that are relevant to this question, referring the reader to Eisenberg and Kauzmann (1969) for details and for discussion of other properties, such as the optical and dielectric properties.

X-ray Diffraction

X-ray diffraction of liquid water provides us with a radial distribution function for water molecules in a coordinate system in which the center of one molecule is the origin. The most accurate measurements are those of Narten et al. (1967). They show a very sharp peak in the distribution function at a distance that increases with increasing temperature, shifting from 2.82 to 2.94 Å as the temperature goes from 4 to 200°C. This peak represents the average distance between nearest-neighbor molecules and indicates, therefore, that nearest neighbors are on the average farther apart in the liquid than they are in ordinary ice, despite the greater overall density of the liquid which, as in the denser polymorphic forms of ice, presumably arises from a decrease in the distance between nonbonded neighbors. The existence of a well-defined nearest-neighbor separation is of course not peculiar to water: a similar parameter can be measured for all liquids. Water, however, differs from most other liquids in the *intensity* of the peak in the distribution function, which tells us how many nearest-neighbor molecules there are. For most simple liquids this number is relatively large (for close-packed spheres there would be 12 nearest neighbors), in water it is only 4.4. This is unequivocal evidence that the local order about a given water molecule remains close to tetrahedral.

At low temperature the radial distribution function contains additional peaks of low intensity at 3.5, 4.5, and 7 Å, but these peaks disappear gradually as the temperature is raised. At distances of 8 Å or above the positions of water molecules have become completely randomized even at 4°C. The separations of 4.5 and 7 Å are consistent with an icelike tetrahedral model, but the separation of 3.5 Å is not. It provides one of the focal points for attempts to define the difference between short-range organization in the crystalline and liquid states.

Thermochemical Data

The heat of sublimation of ice at 0°C is 11,200 cal/mole. Of this amount, 1400 cal/mole represents the translational enthalpy of the vapor, that is, the kinetic energy of the random motion of the molecules in the gaseous state. The remaining 9800 cal/mole represent the energy required to break the bonds that hold the crystal together. This is primarily the energy of the hydrogen bonds. Van der Waals energy, between nonbonded molecules, which makes a major contribution to cohesion in organic crystals, is relatively unimportant here because of the great separation (4.5 Å) between nearest neighbors that are not hydrogen-bonded. The heat of fusion of ice, 1440 cal/mole, is much smaller than the heat of sublimation, and it is thus evident that most of the cohesive energy residing in the hydrogen bonds of the crystalline state is still present in the liquid state. A simplistic interpretation of the numerical results is that about 15% of the hydrogen bonds in ice are broken in the process of fusion. On the other hand, it is equally likely that the hydrogen bonds in the liquid are simply weaker than those in ice (the increase in the average separation of nearest neighbors suggests this), and that the fraction of broken bonds at any instant is very small.

Liquid water, even in the absence of a nonpolar solute, has an anomalous heat capacity, albeit much smaller than those listed in Tables 4-1 and 4-2. C_p is 18 cal/mole-deg in the liquid state and 9 cal/mole-deg both in the crystalline and vapor states. Even this relatively small anomalous heat capacity cannot be explained on the basis of energy states of individual molecules, and must be attributed instead to a temperature dependence of the energy of intermolecular organization. It could represent an increase in the number of broken hydrogen bonds as the temperature is raised, or a shift in equilibrium between two or more discrete configurational states of similar free energy but different enthalpy and entropy, or it could signify that the intermolecular organization is continuously variable, the average state gradually shifting to higher energy as the temperature is raised. It may be noted that the radial distribution function cannot distinguish between these possibilities, since it represents the average positions of molecules over a long time period. The increase in average nearest-neighbor distance at higher temperatures, as noted above, indicates an average weakening of hydrogen bonds, but cannot distinguish between continuous and discontinuous variability in intermolecular distance or energy.

The OH Stretching Vibration

By carrying out infrared or Raman spectral measurements on dilute solutions of HDO in D_2O or H_2O, it is possible to observe the OH or OD stretching vibration in isolation; each is virtually free from the effects of the coupling that occurs between vicinal oscillators of the same fundamental frequency. The vibration frequency is a sensitive function of the O—H···O or O—D···O bond energy and corresponding bond length, and the resolution is high. Frequencies corresponding to icelike hydrogen-bonded OH or OD (measured in ice containing a little HDO) or to nonbonded OH or OD (measured in dilute solutions of HDO in an inert solvent) are represented by narrow bands separated by 300 to 400 cm^{-1}. An important aspect of this measurement is that the process of light absorption occurs on a time scale that is *small* in comparison with the period of molecular displacements leading to the rearrangement of intermolecular organization. One thus obtains an essentially instantaneous signal from each molecule and not an average signal reflecting the various states through which a molecule passes in a given period of time. By this method one can therefore distinguish between a mixture of hydrogen-bonded and nonhydrogen-bonded OH or OD groups, which would be expected to lead to a two-peak absorption spectrum with the relative intensities of the two peaks shifting with temperature, and a continuous range of possible hydrogen-bond energies, which would be expected to lead to a single relatively broad absorption peak lying between the frequencies for icelike and nonhydrogen bonded groups.

Both Raman and infrared spectroscopy (Fig. 5-3) yield broad absorption

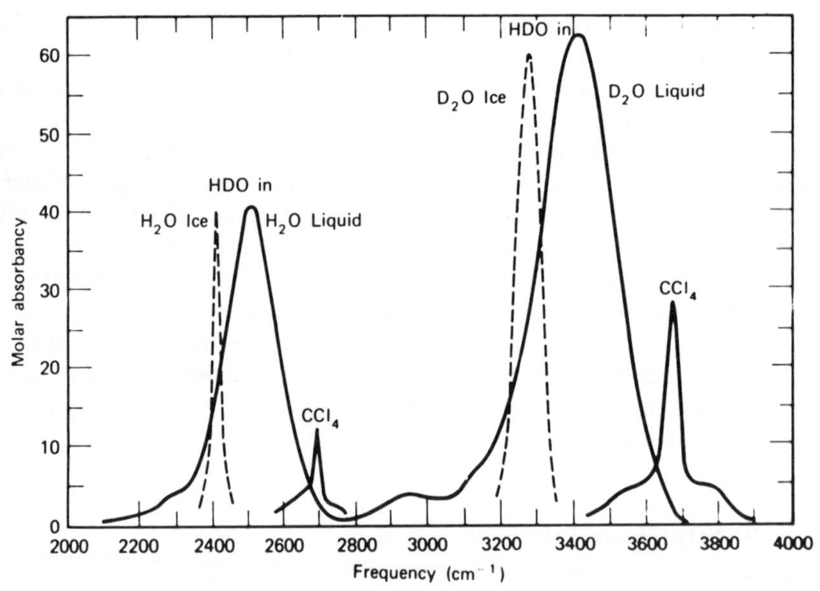

Fig. 5-3. Uncoupled OH and OD stretch bands from infrared spectra of small amounts of HDO in CCl_4, liquid H_2O or D_2O, and crystalline H_2O or D_2O (Stevenson, 1968). From *Structural Chemistry and Molecular Biology* edited by Alexander Rich and Norman Davidson. W. H. Freeman and Company. Cpoyright © 1968. (An additional fact of interest is that the vibrational frequencies in CCl_4 are very close to the OD and OH stretching frequencies of HDO in the vapor state.)

bands of intermediate frequency, with a small increase in peak frequency with temperature (Wall and Hornig, 1965; Stevenson, 1968). This result would seem to be inconsistent with the possible existence in liquid water of a mixture of broken hydrogen bonds and of intact hydrogen bonds of undiminished energy. It provides instead strong evidence for the existence of a broad range of possible hydrogen bond distances. On the basis of the known relation between the vibration frequency and hydrogen bond distance, distribution functions for this distance can be calculated from the spectral data, and they agree well with the time-averaged distribution function derived from X-ray diffraction data (Wall and Hornig, 1965).

It should be pointed out that this interpretation of the spectral data is not universally accepted, and arguments against it have been presented by Frank (1970).

Density

One of the properties of liquid water that has been measured with greatest precision is density (or molar volume) as a function of pressure and temperature. The most striking features are that the density of the liquid is nearly 10% greater than the density of ordinary ice, and that the liquid itself has a negative coefficient of thermal expansion near the melting point, the density increasing to a maximum at 4°C (for H_2O at 1 atm pressure) before the normal positive thermal expansion takes over. D_2O has a similar density maximum at 11.2°C. This intriguing phenomenon turns out not to be a sensitive discriminant between different ideas concerning the liquid structure, because it is primarily a consequence of the open low-density structure that is imposed on ice by the tetrahedral hydrogen-bonded water molecules. Any process that leads to disruption of this structure and increased randomization of the intermolecular arrangement must increase the density. The usual interpretation of the density maximum in the liquid state supposes that this maximum results from a continuation of the same kind of structural rearrangement that distinguishes the liquid state from the crystalline state, with normal thermal expansion acting in opposition.

MODELS FOR LIQUID WATER

The properties of liquid water just discussed and other physical properties do not permit an unequivocal interpretation of the structure of liquid water. The most attractive model is probably the distorted hydrogen-bond model of Pople (1951), recently extended by Bernal (1964), and this is the model favored by Eisenberg and Kauzmann (1969). It is conceptually a very simple model. It assumes that all water molecules in the liquid continue to be hydrogen-bonded to four neighboring water molecules, but that the intermolecular links can be bent and stretched to produce irregular and varied networks. We already know that such distorted networks exist from our examination of the crystal structures of the denser forms of ice and the crystalline gas hydrates, whereas there are no known crystal structures containing broken hydrogen bonds. Distorted and weakened hydrogen bonds are consistent with the spectroscopic data, which seem to exclude the possibility of a significant fraction of broken hydrogen bonds. The model originally formulated by Pople (1951) does not account for the 3.5 Å peak in the radial distribution function obtained by X-ray scattering, but the extension by Bernal (1964) can do so. Second nearest-neighbor water molecules, not hydrogen-bonded to the central water molecule of the coordinate system, lie at a distance of about 3.5 Å from the central molecule in several of the high density crystalline forms of ice.

The distorted bond model for liquid water has thus far not been developed in detail. If accurate information were available for the energy of hydrogen bonds as a function of the extent of distortion, one could in principle evaluate all the properties of liquid water on the basis of this model without arbitrary assumptions. However, this information is not available, and the statistical problem involved in making the requisite calculations of observable properties would in any event be a formidable one.

Many workers who have attempted to account for the properties of liquid water in terms of a structural model have done so in terms of mixture models. There exist at least a dozen such models, all of which share the basic premise that liquid water is a mixture of several distinct molecular species. All of these models require the existence of broken hydrogen bonds: in some models they occur at the boundaries of postulated molecular clusters of different kinds, and in others they arise through the postulated presence of free gas like H_2O molecules in the interstices of hydrogen-bonded networks. By assigning appropriate energy and density parameters to the different molecular species, one is able to evaluate the equilibrium distribution among the several species as a function of external conditions, and to calculate the overall thermochemical properties and the density as a function of, for example, temperature. The claim to validity of a particular model then rests on the often close agreement between such calculations and the corresponding experimental data. Many of the parameters that enter into such a calculation are, however, unknown, and are arbitrarily adjusted to bring this close agreement about. In some instances the number of adjustable parameters equals or exceeds the number of parameters in a purely empirical equation that can be used to fit the data without reference to any structural considerations at all. Under these circumstances the claim to validity based on agreement between theory and experiment loses most of its force, and is not able to counterbalance the principal experimental arguments against mixture theories, which are (a) that they have no counterpart among the known structures of the various forms of crystalline ice, and (b) that they seem to be in conflict with spectroscopic data. Nevertheless, mixture theories have enthusiastic advocates, and Frank (1970), for example, has recently presented a vigorous defense of them against the arguments raised by infrared and Raman spectral data.

One of the most highly developed theories of the structure of liquid water is the mixture theory of Némethy and Scheraga (1962). It has the advantage of great simplicity: variations in configurational energy are considered as arising entirely from the *breaking* of hydrogen bonds. In attempting to fit this theory to experimental thermochemical data, the authors were compelled to adjust their parameters in order to produce what is surely an unacceptably large fraction of broken hydrogen bonds (greater than 50%) and a remarkably low value for the energy required to break a hydrogen bond, which turns out to be only 1.3 kcal/mole. How small this figure is may be seen from the estimate made by Scheraga (1961) in another place, that the energy of breaking a hydrogen bond should normally be about 6 kcal/mole. The figure of 1.3 kcal/mole is in fact closer to the energy required to *distort* a hydrogen bond (a few tenths of a kcal/mole to judge from the enthalpies of transition between various crystalline forms of ice) than to the energy required to *break* a hydrogen bond. One way to look at this result is to consider it as additional evidence in favor of the distorted bond model.

The theory of Némethy and Scheraga (1962) has been extended to reproduce reasonably well the experimental thermodynamic data for solutions of hydrocarbons in water. We have seen that these data unequivocally demonstrate that the water structure near the solute has to be more highly ordered than it is in pure water. Within the strictures of the Némethy–Scheraga model this means that the proportion of hydrogen-bonded clusters must increase, and the fraction of broken hydrogen bonds must decrease. The theory predicts this result, but only by introducing three new adjustable energy parameters, in addition to those already used in formulating the theory for pure water.

STRUCTURAL CHANGE ASSOCIATED WITH THE HYDROPHOBIC EFFECT

In the absence of a convincing working model for the structure of pure liquid water, it is clearly impossible to define the structural changes that occur when hydrocarbon chains are dissolved in water. However, the crystal structures briefly discussed in this chapter provide a frame of reference that may be used as a point of departure. They suggest that water molecules are readily ordered into networks forming cagelike cavities within which nonpolar solutes may be enclosed, and that this may be done without breaking hydrogen bonds or altering the intrinsic coordination pattern of four nearest-neighbor water molecules. The cavity volume is apparently readily variable. The simplest gas hydrates (Fig. 5-2) have dodecahedral cavities large enough to contain Xe or CH_4, and somewhat larger tetrakaidecahedral cavities. In the crystalline hydrate of tetra(isoamyl) ammonium fluoride (Feil and Jeffrey, 1961) an even larger pentakaidecahedral cavity is observed. There is no reason to believe that these two known crystal structures exhaust the possible size range.

If such structures exist around nonpolar molecules in the liquid state, they are likely to be imperfect: random distortions leading to disorder are likely to be superimposed on the uniquely ordered bond distortions that lead to formation of polyhedral cages. The likelihood of such a process is suggested first by analogy: randomness certainly occurs in pure liquid water, since the time-average order as given by the radial distribution function extends to only 8 Å. Second, however, it must be kept in mind that an ordered array of water molecules around a nonpolar solute in a dilute solution is itself surrounded by pure liquid water, and that most water molecules in the postulated polyhedral cages will be hydrogen-bonded to an external water molecule as well as to each other. The ordered structure is thus likely to be influenced by the rapidly fluctuating structural changes around it.

The likely occurrence of imperfections and fluctuations makes it not unreasonable that a hydrocarbon chain that is part of an amphiphilic molecule, and thus unable to generate a completely closed solvent cage, should nevertheless have thermodynamic properties similar to those of a complete hydrocarbon molecule in water solution. One can also readily visualize within the range of possibilities suggested by these considerations why the anomalous heat capacity associated with the hydrophobic effect is so large, and why this heat capacity and the partial molal entropy and enthalpy are not as regular functions of the size of the solute as the free energy is.

REFERENCES

Bernal, J. D. (1964). *Proc. Roy. Soc.*, **A280,** 299.

Eisenberg, D., and W. Kauzmann. (1969). *The Structure and Properties of Water*. Oxford University Press.

Feil, D., and G. A. Jeffrey. (1961). *J. Chem. Phys.*, **35,** 1863.

Frank, H. S. (1970). *Science*, **169,** 635.

Kamb, B. (1968). In *Structural Chemistry and Molecular Biology*. A. Rich and N. Davidson, Eds., W. H. Freeman and Co., San Francisco.

Narten, A. H., M. D. Danford, and H. A. Levy. (1967). *Discuss. Faraday Soc.*, **43,** 97.

Némethy, G., and H. A. Scheraga. (1962). *J. Chem. Phys.*, **36,** 3382, 3401.

Pauling, L. (1935). *J. Am. Chem. Soc.*, **57,** 2680.

Pauling, L. (1960). *The Nature of the Chemical Bond*, 3rd ed., Cornell University Press, Ithaca, N.Y., Chapter 12.

Pauling, L., and E. Marsh. (1952). *Proc. Nat. Acad. Sci. U.S.A.*, **38,** 112.

Pople, J. A. (1951). *Proc. Roy. Soc.*, **A205,** 163.

Scheraga, H. A. (1961). *Protein Structure*, Academic Press, New York.

Stevenson, D. P. (1968). In *Structural Chemistry and Molecular Biology*, A. Rich and N. Davidson, Eds., W. H. Freeman and Co., San Francisco.

Wall, T. T. , and D. F. Hornig. (1965). *J. Chem. Phys.*, **43,** 2079.

SIX

MICELLES

INTRODUCTION

When amphiphilic molecules are dissolved in water they can achieve segregation of their hydrophobic portions from the solvent by self-aggregation. The aggregated products are known as *micelles.*[1] When the hydrophobic portion of the amphiphile is a hydrocarbon chain, the micelles will consist of a hydrocarbon core, with polar groups at the surface serving to maintain solubility in water. Micelles can be small spheres or ellipsoids, or long cylinders, or they can be in the form of bilayers, that is, two parallel layers of amphiphile molecules with the polar groups facing out. Bilayer micelles often form more or less spherical vesicles with an internal solvent-filled cavity.

The hydrocarbon chains in such micelles are generally regarded as disordered, so that the hydrophobic core is in effect a small volume of *liquid* hydrocarbon, differing from a large volume of liquid hydrocarbon only by the restrictive influence of the close proximity of all parts of the liquid to the surface, as a result of which, for example, parts of hydrocarbon chains closest to the polar groups are constrained to be more or less perpendicular to the surface. The effects of such constraints are, however, quite small, at least for micelles formed from simple amphiphile molecules. This can be shown

[1] There is no universally accepted definition of the term "micelle" and some workers would not include aggregates based on a bilayer structure within the definition. The term is used here to designate any water-soluble aggregate, spontaneously and reversibly formed from amphiphile molecules or ions. As will be seen in Chapter 9, the most stable form of such aggregates is dictated by thermodynamic factors, and the conditions under which globular aggregates or bilayer structures will be formed are readily predictable.

both by spectroscopic relaxation methods, which measure the freedom of motion of hydrocarbon chains in the micelle, and by spectroscopic and thermodynamic studies of small solutes dissolved in the micelle, which behave remarkably as they would in a liquid hydrocarbon medium of much greater extent. For some of the more complex amphiphiles a temperature above room temperature is required for the existence of disordered micelles.

The simplest evidence for the liquidlike nature of the micelle interior comes from the ability to dissolve hydrocarbons and other hydrophobic substances within them. One quantitative study leading to precise thermodynamic parameters for this process has been carried out by Wishnia (1963), who measured the solubilities of hydrocarbons in sodium dodecyl sulfate micelles, in 0.1 *M* NaCl as overall solvent. The micelles in this case (from measurements to be cited later) contain on the average about 85 amphiphile ions per micelle (at 25°C) and they are probably ellipsoidal in shape. However, this information is actually not necessary for estimation of thermodynamic parameters, since the mole fraction (X_{mic}) of solute in the micelles can be obtained by knowing only the total number of dodecyl sulfate ions in micellar form and the added solubility of hydrocarbon (above that in 0.1 *M* NaCl alone) that they produce. Assuming that the hydrocarbon within the micelle constitutes an ideal solution, we can obtain the standard free energy of transfer to the micelle interior in the same way as for other solubility measurements, that is,

$$\mu^{\circ}_{\text{mic}} - \mu^{\circ}_{W} = RT \ln (X_W / X_{\text{mic}}) \tag{6-1}$$

Measurements were made at several temperatures from 15 to 35° and heats and entropies of transfer were obtained as outlined in Chapter 4. Unfortunately the data were not of sufficient precision to permit the evaluation of the change in heat capacity. The results obtained are summarized in Table 6-1. A least-squares treatment of the free energies of transfer leads to the relation (in cal/mole)

$$\mu^{\circ}_{\text{mic}} - \mu^{\circ}_{W} = -1934 - 771\, n_{\text{C}} \tag{6-2}$$

The experimental uncertainty is relatively large, ± 160 cal/mole in the constant term and ± 43 cal/mole in the slope.

These results are to be compared with the corresponding data for transfer to a liquid hydrocarbon, as given by equation 2-3 and in Table 4-1, the latter with signs reversed. The similarity is striking. The free energies of transfer are smaller by about 15%, and this difference stems largely from a slightly smaller entropy of transfer. The enthalpies are nearly the same (subject to an experimental error of the order of 500 cal/mole), and even the change in sign with increasing chain length is duplicated. The small

Table 6-1. Thermodynamics of Transfer from Water to the Interior of a Dodecyl Sulfate Micelle[a]

	$\mu^{\circ}_{\mathrm{mic}} - \mu^{\circ}_{W}$ (cal/mole)	$\bar{H}^{\circ}_{\mathrm{mic}} - \bar{H}^{\circ}_{W}$ (cal/mole)	$\bar{S}^{\circ}_{\mathrm{mic}} - \bar{S}^{\circ}_{W}$ (cal/mole)
Ethane	−3450	+2000	+18.3
Propane	−4230	+1000	+17.5
Butane	−5130	0	+17.2
Pentane	−5720	−1100	+15.6

[a] From Wishnia (1963). Results are for 25°C.

difference in entropy undoubtedly reflects the previously mentioned constraints within the micelle that result from the necessity for some degree of orientation near the surface, and the difference is actually surprisingly small. It is clear from these data that the core of a dodecyl sulfate micelle must represent a medium that is very close indeed to a liquid hydrocarbon.

SPECTROSCOPIC RELAXATION METHODS

Spectroscopic relaxation methods have proved to be very useful for estimation of fluidity in micelles and other systems, in terms of the rate of motion of molecules within the systems. All such methods must satisfy two criteria: (1) they must be able to resolve spectral differences that arise from the geometrical orientation of the absorbing entity, and (2) the time scale of the measurement must be comparable to the time scale of the molecular motion one wishes to observe. Free rotational motion of small molecules in a medium of low viscosity occurs with a relaxation time of the order of 10^{-8} sec, and this therefore defines the required time scale of the experiment. Relaxation processes for larger molecules or for particles such as micelles occur more slowly, and one can therefore distinguish between motion of a molecule (or a portion of a molecule) within the micelle and motion of the micelle as a whole. An increase in the viscosity of the medium also slows down the relaxation process, and formation of an ordered phase will abolish some motions entirely, so that the state of the local environment near the molecule that is being observed can be investigated.

One method satisfying these criteria is *fluorescence depolarization* (for an introduction with reference to basic source material, see Stryer, 1968). The lifetime of the excited state in fluorescence measurements is of the order of nanoseconds, and dependence on geometrical orientation can be readily introduced by using polarized light for excitation, so that in a liquid medium chromophores with transition moments parallel to the plane of polarization will be preferentially excited. In the absence of molecular motion, the emitted fluorescent radiation will be partially polarized, reflecting the preferred orientation of the emitters. Molecular motion within the lifetime of the fluorescent state will result in diminution of the extent of polarization. The most straightforward procedure is to use a pulse of nanosecond duration for excitation and then to follow the polarization of the fluorescence as a function of time for as long as it can be observed. If the molecular motion is isotropic (no preferred axis for rotation) it is characterized by a single relaxation time τ, with the result that the degree of polarization at time t, $P(t)$ is given by

$$P(t) = P(0)e^{-3t/\tau} \tag{6-3}$$

where $P(0)$ is the degree of polarization at some suitably defined zero time. If the motion is anisotropic there would be more than one exponential term on the right-hand side of equation 6-3. A less direct method is to use steady state measurements, that is, to measure the polarization of fluorescence under constant excitation. This is essentially equivalent to measurement of the molecular motion at a single time determined by the lifetime of the excited state, and requires assumptions about whether the motion is isotropic or not. If equation 6-3 is deemed applicable, $P(0)$ must be determined independently, and this is normally done by varying the viscosity of the medium and extrapolating to infinite viscosity, a procedure that may not always be valid.

Since hydrocarbon chains are not fluorescent at usable wavelengths, the micelle interior may be investigated by use of *fluorescent probes*, which are molecules containing suitable chromophores that are soluble in the micelle. That a given probe is actually dissolved in the hydrophobic core can be readily established since the characteristic absorption and emission spectra are usually significantly affected by the polarity of the medium: some, like the frequently employed probe 1-anilino-8-naphthalene sulfonate, display virtually no fluorescence in water. Shinitzky et al. (1971) have used two aromatic hydrocarbon probes of this kind to investigate micelles formed by several cationic amphiphiles, employing the steady state method. The relaxation times observed were found to be too fast to be ascribable to motion of the micelle as a whole and depended little on micelle size. They were somewhat slower, however, than relaxation times observed for the same chromophores in pure liquid hydrocarbons of appropriate chain length. Both from the relaxation times themselves and from their dependence on temperature, they concluded that the micelle interiors are liquid, albeit with a higher viscosity than a pure hydrocarbon such as dodecane would exhibit.[2] This higher viscosity is to be expected, since the hydrocarbon chains of micelles are anchored to the surface hydrophilic groups, and thus restricted in their motility. The same effect is seen in many other systems: for example, liquid aliphatic alcohols have a viscosity several times higher than the corresponding liquid hydrocarbons.

A second procedure is to use the election spin resonance (esr) spectrum observable for free radicals in a magnetic field (Hamilton and McConnell, 1968; Griffith and Waggoner, 1969). Since the systems of interest do not contain free radicals, the use of a probe substance is again required. The dialkyl nitroxide radical

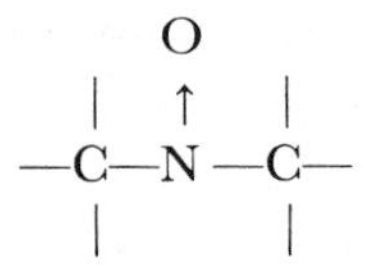

has been found to be especially useful, since it is remarkably stable and unreactive, and can be readily incorporated into a variety of organic molecules. The resonance time is again of the order of 10^{-8} sec, and the geometrical factor here is provided by the unique direction associated with the spin transition: the hyperfine splitting seen in the esr spectrum is resolvable into two components ($T_{\parallel}$ and $T_{\perp}$), one corresponding to the transition moment parallel to the applied magnetic field, the other to perpendicular orientation. The characteristic positions and intensities for a given probe can be determined by use of crystalline or otherwise ordered systems. Molecular motions that occur more rapidly than 10^{-8} sec result in sharp resonance lines that represent the *time average* of all geometric states encountered during the resonance period. In a liquid, where motion is completely isotropic, only a single value for T is therefore observed. On the other hand, if interchange of geometrical states is slow, both components will be observed. Rapid *anisotropic* motion (e.g., torsional motion about a fixed molecular axis) results in observation of both components, but with characteristics different from those of a crystalline sample. In that case the difference between the observed values for $T_{\parallel}$ and $T_{\perp}$ and those in the crystalline reference can be used to

[2] Rotational diffusion constants are inversely proportional to the viscosity of the medium, and an apparent viscosity (termed *microviscosity* by Shinitzky et al. (1971)) can be obtained either by measuring τ for the same probe in liquids of different viscosities or, if the size and shape of the probe molecule can be established, by the use of equations relating size, shape, and solvent viscosity. Shinitzky et al. obtained microviscosities of 17 to 32 CP at 27° for the micelle interior, when using amphiphiles with alipathic hydrocarbon chains of 12 to 16 carbon atoms. These values are to be compared with viscosities of 3.0 CP for liquid *n*-hexadecane at a comparable temperature, and of 13.4 CP for *n*-hexadecyl alcohol at 50°C (about twice this value is to be expected at 25°C).

determine the average orientation resulting from the rapid motion, since there is a simple relation between the observed value of T and the contributions of the reference values to it. Line width and shape are also helpful in the analysis of local modes of motion, as in nuclear magnetic resonance (see below).

Nitroxide free radical probes introduced into sodium dodecyl sulfate micelles do not show separate components in the esr spectrum, thus indicating that they are in a liquid medium, undergoing isotropic motion (Waggoner et al. 1967). The spectral lines are somewhat broader, however, than they are in aqueous solution indicating, as did the fluorescence measurements discussed previously, that the local viscosity is fairly high. More extensive use of the technique has been made for micelles formed by biological lipids, and these results are discussed in Chapter 13. These results and inferences drawn from them suggest that the segments of the hydrocarbon chain nearest the amphiphile head group will generally (i.e., in micelles formed by simple amphiphiles, as well as those formed by biological lipids) be more restricted in their motion than more distal parts of the chain, as is to be expected.

A disadvantage of both fluorescence depolarization and esr spectroscopy is that they require the use of an external probe. It is possible that the hydrocarbon chains in the vicinity of the probe are not in the same state as they would be in a micelle containing no additives, although there is actually good evidence (some of which will be mentioned later in the book) that no important perturbations are produced. An experimental tool that avoids the use of an external probe is nuclear magnetic resonance, which can be used to investigate the motion of constituent nuclei of the hydrocarbon chain directly. Most investigators have used proton magnetic resonance (pmr), in which one is looking at the motion of hydrogen nuclei, but apparatus for observing C atoms (using the paramagnetic isotope C^{13}) has begun to be used recently. A major *disadvantage* of these methods is the absence of an intrinsic geometric factor in the transition moment, so that the rate of motion of elements of the hydrocarbon chain affects only secondary perturbations of the signal, that is, those produced by spin–spin or spin-lattice coupling. The interactions involved are often too numerous to be resolved. The method can always give a qualitative result: if all motions are rapid enough to permit time-averaging of all such interactions within the resonance period (again about 10^{-8} sec) a sharp absorption line will be obtained; if motion is slow or restricted, line broadening will be observed, sometimes to the extent that a line will not be visible at all against background noise. A quantitative interpretation of line broadening in intermediate situations is usually difficult. Waggoner et al. (1967) have shown that line widths in the pmr spectrum of the CH_2 and CH_3 groups of dodecyl sulfate micelles in water are as narrow as those observed in liquid dodecanol, showing that their freedom of motion is also the same. They also dissolved a nitroxide free radical probe in the micelles and, in addition to obtaining the esr spectrum of the probe itself, as discussed above, determined its effect on the line width of the CH_2 signal in the pmr spectrum. A broadening is observed, arising from the added spin–spin interaction, and it was found to be quantitatively consistent with a calculation based on free rapid motion of the probe in the micelle interior.

CRYSTALLIZATION OF VERY LONG HYDROCARBON CHAINS

The foregoing data have shown that amphiphiles with ionic head groups and hydrocarbon tails of up to 16 carbon atoms form micelles with essentially liquid cores at room temperature. It is important to note in this connection that pure saturated hydrocarbons form crystalline solids that melt below room temperature only when $n_C < 18$, as shown by the data of Table 6-2.

Table 6-2. Thermodynamics of Melting of Crystalline Hydrocarbons[a]

	Melting Point (°C)	$\Delta H°$ (kcal/mole)	$\Delta S°$ (cal/deg mole)	$\Delta G°$ at 25°C (kcal/mole)
$C_{13}H_{28}$	− 5.4	6.8	25.4	−0.77
$C_{14}H_{30}$	5.9	10.8	38.6	−0.74
$C_{15}H_{32}$	10.0	8.3	29.2	−0.44
$C_{16}H_{34}$	18.2	12.8	43.8	−0.30
$C_{17}H_{36}$	22.0	9.7	32.8	−0.10
$C_{18}H_{38}$	28.2	14.8	49.2	+0.16
$C_{19}H_{40}$	32.0	12.0	36.0	+0.25
$C_{20}H_{42}$	36.7	16.8	54.2	+0.63

[a] Data are for saturated *n*-paraffins based on compilations by the American Petroleum Institute (1953). Values of $\Delta H°$ and $\Delta S°$ refer to the melting point. Values of $\Delta G°$ at 25°C are calculated assuming constant $\Delta H°$ and $\Delta S°$ between 25°C and the melting point. Crystals formed by unsaturated hydrocarbons melt at much lower temperatures.

The crystal structure is based on parallel extended hydrocarbon chains, and the packing differs between molecules with odd and even values of n_C, as is reflected in the alternation of thermodynamic properties seen in Table 6-2. Because hydrocarbon chains are necessarily bent at an angle where double bonds occur, unsaturated hydrocarbons form less stable crystals than saturated ones, which melt well below room temperature for all chain lengths of interest.

These data suggest that micelles formed by amphiphiles with very long saturated hydrocarbon chains may have an ordered hydrocarbon core at room temperature, which becomes liquidlike only at higher temperatures. Moreover, where head group interactions favor ordering of the hydrocarbon chains, the transition from ordered to liquidlike core may occur at higher temperatures than the melting points of pure hydrocarbons of comparable chain length. Micelles formed from amphiphiles containing unsaturated hydrocarbon chains would be much less likely to contain ordered structures.

The following chapters will consider the thermodynamics and other general principles of micelle formation. The discussion will be confined to micelles with liquid cores. Experimental data bearing on ordered arrangements of hydrocarbon chains will be taken up in Chapters 12 and 13.

THE PRINCIPLE OF OPPOSING FORCES

Micelles formed by simple amphiphiles are often quite small, with an aggregation number of the order of 100. Their formation clearly requires the existence of two opposing forces, an attractive force favoring aggregation and a repulsive force that prevents growth of the aggregates to large size. Even in systems where much larger micelles are formed a repulsive force must be present, to prevent separation of the amphiphile into an entirely distinct phase.

For micelles in aqueous solution, such as we are considering, the attractive force arises from the hydrophobic effect acting upon the hydrocarbon chains of the amphiphile. An important feature is that a minimum number of amphiphiles have to become associated with each other before an effective elimination of the hydrocarbon–water interface can be achieved. Two or three amphiphiles cannot form a stable micelle, regardless of whether the hydrocarbon core is to be liquid or ordered. Thus micelle formation is necessarily a cooperative process, requiring simultaneous participation by many amphiphilic molecules or ions. The micelles formed cannot have a purely statistical size distribution, but are limited by this factor in how small they can be, in addition to the limitation on large size that may be imposed by the repulsive force.

The repulsive force in micelle formation must come primarily from the head groups. In ionic micelles electrostatic repulsion between like charges is a major factor. In micelles formed by amphiphiles with uncharged head groups a preference for hydration, as opposed to self-association, is involved. Micelles are formed by amphiphiles with a polyoxyethylene head group. —$(OCH_2CH_2)_nOH$, or with carbohydrate head groups. They are *not* formed by simple aliphatic alcohols, which prefer to associate with each other as a pure liquid, as shown by the data of Fig. 3-1.

REFERENCES

American Petroleum Institute (1953). *Selected Values of Physical and Thermodynamic Properties of Hydrocarbons and Related Compounds.* Carnegie Press, Pittsburgh.

Griffith, O. H., and A. S. Waggoner. (1969). *Acc. Chem. Res.*, **1,** 17.

Hamilton, C. L., and H. M. McConnell. (1968). In *Structural Chemistry and Molecular Biology*, A. Rich and N. Davidson, Eds., W. H. Freeman and Co., San Francisco, p. 115.

Shinitzky, M., A. C. Dianoux, C. Gitler, and G. Weber. (1971). *Biochemistry*, **10,** 2106.

Stryer, L. (1968). *Science*, **162,** 526.

Waggoner, A. S., O. H. Griffith, and C. R. Christensen. (1967). *Proc. Nat. Acad. Sci., U.S.A.*, **57,** 1198.

Wishnia, A. (1963). *J. Phys. Chem.*, **67,** 2079.

SEVEN

THERMODYNAMICS OF MICELLE FORMATION

The literature on the thermodynamics of micelle formation is voluminous, and can be very confusing. (For reviews, see Shinoda et al., 1963; Becher, 1967; Hall and Pethica, 1967; Mukerjee, 1967; Anacker, 1970.) The most detailed and rigorous analysis is that of Hall and Pethica (1967), which is based on the general thermodynamic theory for systems of this type by Hill (1964). This analysis is the definitive reference for exact thermodynamic relationships. However, it is not possible to gain physical insight into micelle formation without dividing the free energy into separate attractive and repulsive contributions, and this separation cannot be treated in terms of rigorous classical thermodynamics. The rigorous treatment is therefore of limited utility in satisfying the objectives of the present chapter.

THERMODYNAMIC FORMULATION

The thermodynamic parameter that determines the equilibrium between micelles and free amphiphile in solution, and the distribution between micelles of different size, is the difference between the standard unitary free energy of an amphiphile molecule or ion in a micellar aggregate, which we shall call μ°_{mic}, and the corresponding quantity for the free amphiphile in aqueous solution, μ°_{W}. It is important to keep in mind that μ°_{mic} and other quantities with the subscript "mic" used in this chapter refer to individual amphiphile molecules in a micellar aggregate and not to the chemical potential of the micellar aggregate as a whole.

To relate thermodynamic parameters for micelle formation to experimental data, it is sometimes assumed that the micelles can be treated as a separate phase. In that case the Gibbs phase rule would apply, and micelles and free amphiphile could be in equilibrium at a given temperature and pressure only at a single fixed value of X_W, which would effectively be a "solubility." Thus the treatment of Chapter 2 would apply, with

$$\mu^{\circ}_{\text{mic}} - \mu^{\circ}_{W} = RT \ln X_W + RT \ln f_W \tag{7-1}$$

This equation is, however, incorrect. Micelles are in fact dispersed in solution, and equation 7-1 neglects the cratic contribution to μ_{mic} that arises from the mixing of micelles with solvent. To incorporate this contribution the micelles have to be treated as a component of the homogeneous aqueous phase, and the phase rule restriction on the free amphiphile concentration does not apply.[1]

To obtain the correct equilibrium relation several alternative procedures are possible.

1. All micelles can be considered to be a single component of the solution, regardless of micelle size. In that case the cratic contribution to the free energy *per mole of micelles* would be $RT \ln$ (mole fraction of micelles), assuming that one is dealing with a solution sufficiently dilute in micelles to permit neglect of nonideality terms arising from interaction of micelles with each other. If the micelles contain on the average $\bar{m}$ amphiphile molecules, the cratic contribution per mole of amphiphile would be $1/\bar{m}$ of the contribution per mole of micelles. The mole fraction of amphiphile incorporated into micelles (X_{mic}) would be $\bar{m}$ times the mole fraction of micelles. Thus

$$\mu_{\text{mic}} = \mu^{\circ}_{\text{mic}} + \frac{RT}{\bar{m}} \ln \frac{X_{\text{mic}}}{\bar{m}} \tag{7-2}$$

[1] In fact, to consider micelles as a separate thermodynamic entity at all is purely a matter of convenience. See Hill (1964), page 120. It is conceptually useful to do so because in most instances a minimum number of amphiphile molecules must become associated with each other before a micelle with a liquid hydrocarbon core can be said to exist, and micelles with m smaller than this number are therefore not thermodynamically stable. As a result, micelle size distribution functions tend to be bellshaped, as illustrated by Fig. 8-5, and "micelles" tend to be distinct from "free" amphiphile molecules. This is true even if the monomeric amphiphile molecule has some tendency to form dimers, as has been suggested for some systems: the dimers would be considered as belonging to the class of "free" amphiphile molecules and would be allowed for thermodynamically in terms of an activity coefficient in the equation for the chemical potential of the non-aggregated amphiphile.

It is probable that the concept of micelles would not be useful if micelle formation were the result of noncooperative random polymerization (Tanford, 1961; chapter 3). In that case the size distribution function would show a continuous decrease in the number of aggregates from $m = 1$ to $m = \infty$.

At equilibrium, $\mu_{\rm mic}$ must be equal to the chemical potential of free amphiphile in solution, as given by equation 2-1, so that we obtain, in place of equation 7-1,

$$\mathring{\mu}_{\rm mic} - \mathring{\mu}_W = RT \ln X_W + RT \ln f_W - \frac{RT}{\bar{m}} \ln \frac{X_{\rm mic}}{\bar{m}} \tag{7-3}$$

The value of $\mathring{\mu}_{\rm mic}$ given by this equation relates to the probability of an amphiphile being in the micellar state, characterized by a given $\bar{m}$, regardless of how the particles in that state may be distributed with respect to specific values of m. The value of $\mathring{\mu}_{\rm mic} - \mathring{\mu}_W$ given by equation 7-3 is appropriate for comparison with other free energies of transfer, such as $\mathring{\mu}_{HC} - \mathring{\mu}_W$.

2. Micellar aggregates of different size can be considered to be distinct species, each characterized by its own value of the standard chemical potential, $\mathring{\mu}_{{\rm mic},m}$. Since $\mu_{\rm mic}$ must be the same in all micelles at equilibrium, we obtain, in place of equation 7-3,

$$\mathring{\mu}_{{\rm mic},m} - \mathring{\mu}_W = RT \ln X_W + RT \ln f_W - (RT/m) \ln (X_{{\rm mic},m}/m) \tag{7-4}$$

where $X_{{\rm mic},m}$ represents the mole fraction of amphiphile incorporated specifically into micelles containing m amphiphiles. This equation is useful for considering the distribution function for micelle size. Note that $\mathring{\mu}_{\rm mic}$ of equation 7-3 is not an average of the $\mathring{\mu}_{{\rm mic},m}$ of equation 7-4, but more negative than any one of them, since the probability of an amphiphile being in any micelle regardless of size is necessarily greater than the probability of being in a micelle of a particular size.

3. One can express the equilibria in terms of equilibrium constants for micelle formation, that is,

$$K_m = X_{{\rm mic},m}/X_W{}^m f_W{}^m \tag{7-5}$$

as was done, for instance, by Debye (1949). This approach is identical to the preceding one, with $m(\mathring{\mu}_{{\rm mic},m} - \mathring{\mu}_W)$ equal to $-RT \ln K_m$.

CRITICAL MICELLE CONCENTRATION (CMC)

The concept of a "critical concentration" for the formation of micelles from free amphiphile is inexact but convenient. The use of this concept is probably a major cause of confusion in the thermodynamic analysis of micelle-forming systems.

If micelle formation could be regarded as separation of a distinct phase, free amphiphile in solution could coexist with the micellar phase at only a single value of X_W as given by equation 7-1. As shown in Fig. 7-1*a*, this unique value of X_W would be measurable in two ways: either as the total

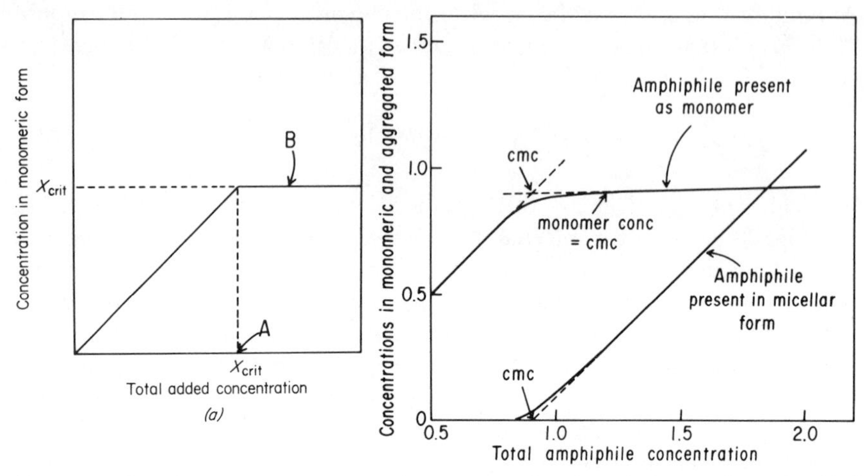

Fig. 7-1. Relation between monomeric concentration in solution and total added concentration in (*a*) true phase separation and (*b*) micelle formation. Figure 7-1*b* is based on equation 7-3 with $\overline{m} = 50$ and $f_W = 1$. Both μ°_{mic} and $\overline{m}$ are assumed independent of concentration within the range covered. The figure shows the concentration of amphiphile present in micellar form as well as the monomer concentration. The dashed lines show empirical procedures for determining the cmc: the point at which the monomer concentration is equal to the cmc is also indicated. Concentration units are arbitrary and could be expressed in grams of amphiphile or in moles based on monomer molecular weight.

added concentration at which phase separation first occurs (point A), or as the free concentration in solution as analytically measured in the presence of the micellar phase (line B).

In fact micelle formation is not equivalent to phase separation. When micelles and free amphiphiles coexist, X_W is a variable, depending on the concentration of micelles present according to equation 7-3 or the set of equations 7-4 for separate values of m. However, if $\overline{m}$ is reasonably large ($\overline{m} \gtrsim 50$) X_W changes only slowly as X_{mic} increases, and the transition from the predominantly unassociated amphiphile to the micellar state does in fact occur over a narrow critical range of concentration, approaching true phase separation in sharpness, as illustrated by Fig. 7-1*b*.

It is customary to define a single concentration within this transition zone as a *critical micelle concentration*, generally abbreviated as "cmc." This is usually done by empirical graphical procedures, such as shown by the dashed lines of Fig. 7-1*b*. In practice, X_W and X_{mic} are not necessarily measured explicitly, and the cmc determination is based on a change in slope when an appropriate physical property that distinguishes between micellar and free amphiphile is plotted against total concentration. It is important to emphasize that no procedure can yield a unique critical concentration because none in fact exists, even in terms of explicit measurement of X_W and X_{mic}. The lines representing X_W and X_{mic} in Fig. 7-1*b* continue to curve very slightly far from the transition zone, and the cmc value depends to some extent on the concentration range from which the extrapolation to the dashed lines of the figure is performed. However, as long as measurements are derived from data reasonably close to the transition zone, the ambiguity in the cmc generally does not exceed 1 or 2%.

An important feature of micelle formation, as previously noted, is that the free amphiphile concentration in equilibrium with micelles changes only slowly with the concentration of micelles. Thus $X_W \simeq$ cmc over a wide range of conditions. In Fig. 7-1*b*, where the empirically established cmc is 0.90 in the arbitrary units that have been used, $X_W = 0.885$ when $X_{\mathrm{mic}} = 0.12$, $X_W = 0.90 =$ cmc when $X_{\mathrm{mic}} = 0.30$, $X_W = 0.92$ when $X_{\mathrm{mic}} = 1.00$, and so on. Within the accuracy with which the cmc can be defined, X_{mic} in equation 7-3 may be set equal to any multiple of X_W in the range 0.01 to $1.0X_W$, and X_W itself may be replaced by the graphically determined cmc value, that is, equation 7-3 may be written as

$$\mathring{\mu}_{\mathrm{mic}} - \mathring{\mu}_W = [(\bar{m} - 1)/\bar{m}]\, RT \ln \mathrm{cmc} + RT \ln f_W + (RT/\bar{m}) \ln (\sigma \bar{m}) \quad (7\text{-}6)$$

where σ may be anywhere between 1 and 100, and the cmc is in mole fraction units. The choice of a value for σ within the given range affects the result to the extent of only ± 25 cal/mole when $\bar{m} = 50$, and by less than that when $\bar{m}$ is larger. For very large micelles the quotient $(\bar{m} - 1)/\bar{m}$ may be set equal to unity and the final term in equation 7-6 may be neglected. This reduces equation 7-6 to

$$\mathring{\mu}_{\mathrm{mic}} - \mathring{\mu}_W = RT \ln \mathrm{cmc} + RT \ln f_W \quad (7\text{-}7)$$

which is the same as equation 7-1 with $X_W =$ cmc. Thus it is seen that treating micelle formation as true phase separation becomes a progressively better approximation as $\bar{m}$ increases.

EXPERIMENTAL RESULTS

Theoretically, the simplest way to analyze the thermodynamics of micelle formation would be to determine the concentration of free amphiphile in equilibrium with micelles over a range of concentration, and then to apply equation 7-3. In practice only a single parameter, the cmc, is measured and, as we have indicated, this parameter is not uniquely definable. The uncertainty generated thereby turns out to be entirely negligible, partly because good experimental values for micelle size, as required in equation 7-3, are not generally available and, also, because our ability to interpret the results is so limited that greater accuracy could not be used at the present time for interpretative purposes. (For example, even the difference between equations 7-8 and 7-9 below, which is outside experimental error, will be seen to be within the uncertainty in our ability to interpret the data.)

Representative experimental data are shown in Fig. 7-2. We shall first consider the results for N-alkyl betaines, $RN(CH_3)_2^+CH_2COO^-$, where R represents a saturated hydrocarbon chain. Accurate $\bar{m}$ values for this system are available (Swarbrick and Daruwala, 1969, 1970), so that equation 7-6 can be employed. Assuming that the free amphiphile exists in monomeric form and that there is no concentration-dependent contribution to the excess chemical potential, we can set $\ln f$ equal to zero. Least-squares analysis of the results then leads to the relation

$$\mu^{\circ}_{\text{mic}} - \mu^{\circ}_{W} = 2931 - 732n_{C} \tag{7-8}$$

with an uncertainty of ± 10 cal/mole in the slope and ± 120 cal/mole in the intercept. If equation 7-7 is used in place of equation 7-6, the result would be

$$\mu^{\circ}_{\text{mic}} - \mu^{\circ}_{W} = 2514 - 709n_{C} \tag{7-9}$$

with about the same experimental uncertainty. This shows that the error in using the phase separation model for micelle formation for the estimation of

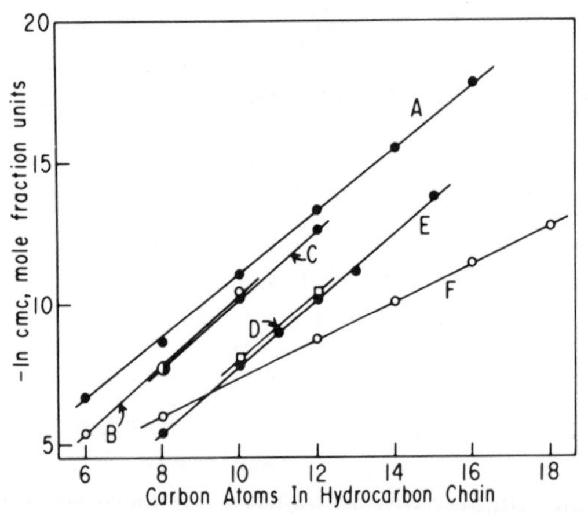

Fig. 7-2. Plots of ln cmc versus hydrocarbon chain length at 25°C, unless otherwise stated. *A*, Alkyl hexaoxyethylene glycol monoethers from Becher (1967). *B*, Alkyl sulfinyl alcohols (Corkill et al., 1966). *C*, Alkyl glucosides from Becher (1967). *D*, Alkyl trimethylammonium bromides in 0.5 *M* NaBr (Emerson and Holtzer, 1967a; Geer et al., 1971). *E*, *N*-alkyl betaines (Swarbrick and Daruwala, 1969). *F*, Alkyl sulfates in the absence of added salt, at 40°C (Evans, 1956).

$\mu^{\circ}_{\text{mic}} - \mu^{\circ}_{W}$ is not severe, even in this system where the micelle size is quite small ($\bar{m}$ varies from 24 for $n_C = 8$ to 130 for $n_C = 15$). In systems where the micelle size is larger, the error would be even less.

As Fig. 7-2 shows, experimental cmc values for other micelles have somewhat different absolute values from those for the N-alkyl betaines, but, with the exception of ionic micelles in the absence of added salt, have very nearly the same dependence on hydrocarbon chain length. This is reflected in the $\mu^{\circ}_{\text{mic}} - \mu^{\circ}_{W}$ values determined from the data: the constant term varies, depending on the nature of the head group, but the dependence on n_C does not. For the alkyl hexaoxyethylene glycol monoethers, for example, using equation 7-6

$$\mu^{\circ}_{\text{mic}} - \mu^{\circ}_{W} = 715 - 708n_C \tag{7-10}$$

with an uncertainty of ±24 cal/mole in the slope and ±290 cal/mole in the intercept. The uncertainty is relatively large because the micelle aggregation numbers (Balmbra et al., 1964) are difficult to determine in this system and are not known with great accuracy. If the phase separation model (equation 7-7) is used

$$\mu^{\circ}_{\text{mic}} - \mu^{\circ}_{W} = 257 - 687n_C \tag{7-11}$$

and the precision is much better: ± 8 cal/mole in the slope and ± 90 cal/mole in the constant term.

Both the constant and n_C-dependent terms in these relations merit discussion. The increment per carbon atom in all of them is smaller than similar figures cited previously for other processes involving amphiphile molecules: for transfer of an amphiphile from water to a hydrocarbon or alcohol solvent the corresponding parameter has a value of − 820 cal/mole/C atom, for transfer of a hydrocarbon to the interior of a dodecyl sulfate micelle it is −770 cal/mole/C atom (equations 3-2, 3-3, and 6-2). The constant terms in the equations are all positive, as is true also for the free energy of transfer of amphiphiles to organic solvents (equations 3-2 and 3-3), but the same reason cannot apply since hydrophilic head groups are not transferred into the nonaqueous medium in micelle formation. Both aspects of the data can be qualitatively explained in terms of the opposing forces that are involved in micelle formation, as will be shown below.

When ionic micelles are studied in the absence of added salt, a much smaller dependence of the cmc on chain length is observed, as is illustrated by the data for alkyl sulfates in Fig. 7-2. The increment in $\mu^{\circ}_{\text{mic}} - \mu^{\circ}_{W}$ per added carbon atom is only −418 cal/mole, as compared to the value of about −700 cal/mole applicable to the other data in the figure. Slopes of similar plots for several other ionic amphiphiles, in the absence of added salt, also lie close to −420 cal/mole (Shinoda et al., 1963). These greatly reduced slopes will also be seen to be consistent with expectations based on the principle of opposing forces.

There exist a large number of cmc measurements of ionic micelles in the absence of added salt, and some interesting observations may be based on them. Klevens (1953), for example, has found that the cmc's of alkyl carboxylates, alkyl ammonium chlorides, and alkyl sulfonates, for the same length of hydrocarbon tail, are nearly the same, but that the alkyl sulfates of a given chain length have cmc's comparable to those of the other amphiphiles with an additional carbon atom. In other words, the extra oxygen atom in the sulfate head group ($—O—SO_3^-$) behaves as if it were an extra CH_2 group. This finding suggests that the curvature at the surface of a small ionic micelle and the repulsion between head groups may cause a part of the hydrocarbon tail close to the ionic site to remain in an aqueous environment, and to contribute nothing to $\overset{\circ}{\mu}_{\text{mic}} - \overset{\circ}{\mu}_W$. In this situation an oxygen atom (which is spatially equivalent to a CH_2 group) could be substituted for a CH_2 group near the ionic site without affecting the cmc. The very limited data available at high ionic strength indicate that Klevens' generalization is not confined to cmc's in the absence of added salt. For example, the cmc of dodecyl sulfate in 0.5 *M* NaCl is virtually the same as the cmc of C_{13} trimethyl ammonium bromide in 0.5 *M* NaBr, as obtained by extrapolation of the two points for homologous amphiphiles in Fig. 7-2. (It might be noted that other data, not cited here, indicate that a phenyl group inserted into an aliphatic hydrocarbon chain is roughly equivalent to 3 CH_2 groups.)

The effect of unsaturation on $\overset{\circ}{\mu}_{\text{mic}} - \overset{\circ}{\mu}_W$ has been studied by use of ionic micelles in the absence of added salt, and it appears to be similar to its effect on $\overset{\circ}{\mu}_{HC} - \overset{\circ}{\mu}_W$ as measured by hydrocarbon solubility. Comparison between cmc values for potassium oleate and potassium stearate, and for potassium abietate and potassium dehydroabietate (Klevens, 1953) shows that introduction of a double bond (in these instances near the middle of the hydrocarbon chain) has an effect equivalent to removal of 1 to 1.5 carbon atoms from a saturated fatty acid chain.

An interesting study by Evans (1956) makes it possible to assess the effect of having two separate hydrocarbon chains on an amphiphile molecule with a single head group. Evans measured the cmc of compounds of the type

$$
\begin{array}{l}
R_1—CH—OSO_3^- \\
\qquad\;\; | \\
\qquad\;\; R_2
\end{array}
$$

and some of his results are summarized in Fig. 7-3. The results demonstrate unequivocally that a second hydrocarbon tail, added to an amphiphile molecule already possessing a longer tail, makes a smaller contribution to

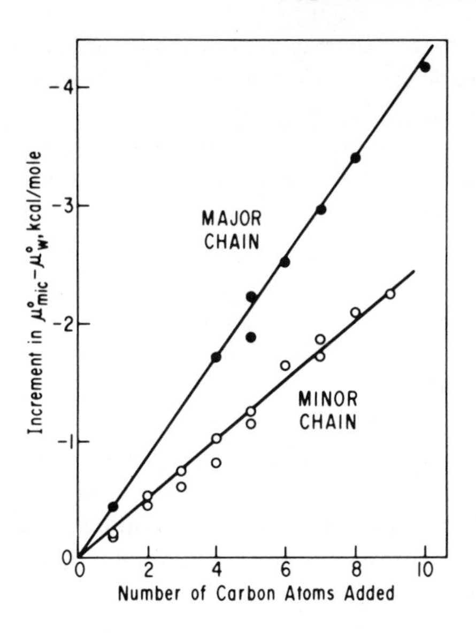

Fig. 7-3. Increment in $\mu^{\circ}_{\text{mic}} - \mu^{\circ}_W$ on lengthening the hydrocarbon chains of 1′-(alkyl)′-alkyl-1-sulfates. Data of Evans (1956), at 40°C in the absence of added salt. Major chain data are for single-chain alkyl sulfates with octyl sulfate as reference and for double-chain derivatives with 1-butyl-decyl sulfate and 1-pentyl-hexyl sulfate as reference. Minor chains data represent a variety of compounds, the parent major chain in each case being the reference compound.

$\mu^{\circ}_{\text{mic}} - \mu^{\circ}_W$ than is made by the longer hydrocarbon chain. On the other hand, the increment per carbon atom added to the longer chain is not affected by the presence of a shorter chain. These results suggest that two hydrocarbon chains on the same amphiphile molecule (in the monomeric state in aqueous solution) tend to associate with each other so that the hydrocarbon-water interface is reduced.

Similar studies with dialkyl dimethylammonium halides (Shinoda et al., 1963, p. 46) lend partial, but not complete support for this conclusion. The cmc of the dioctyl compound in this series is about the same as that of the single-chain undecyl compound. On the other hand, the cmc of the didecyl compound is only slightly higher than that of the compound containing one octyl and one dodecyl chain.

APPLICATION OF THE PRINCIPLE OF OPPOSING FORCES

It was pointed out in the previous chapter that soluble micelles can be formed only if growth is limited by a repulsive force. It is thus realistic to consider μ°_{mic} as composed of two factors,

$$\mu^{\circ}_{\text{mic}} = U^{\circ}_{\text{mic}} + W^{\circ}_{\text{mic}} \tag{7-12}$$

U°_{mic}, which will be negative in sign relative to μ°_W, represents the contribution of the attractive force, and W°_{mic}, which will be positive, represents the repulsive force. Exact calculations of these separate contributions are not possible, but rough estimates can be made to provide a semiquantitative explanation for some of the results cited above.

For an approximate estimate it is probably reasonable to consider $U^\circ_{mic} - \mu^\circ_W$ as independent of micelle size and closely related to the free energy of transfer of a hydrocarbon chain from water to a liquid hydrocarbon, as given by equation 2-4 or by the contribution of the alkyl chain to equation 3-2; or to the somewhat smaller free energy of transfer to the inside of a micelle, as given by the somewhat less accurate data of equation 6-2. Thus the terminal methyl group of an amphiphile alkyl chain would be expected to contribute about $-$ 2000 cal/mole to $U^\circ_{mic} - \mu^\circ_W$, and most of the CH_2 groups would contribute about -800 cal/mole each. However, the CH_2 group closest to the head group is not hydrophobic at all (page 12) and will make no contribution. For small globular micelles an additional one or two CH_2 groups may be outside the micelle core for the reasons given on page 52 in the discussion of the difference between alkyl sulfates and similar ionic amphiphiles. Finally CH_2 groups closest to the surface of the core are likely to be constrained to be more or less perpendicular to the surface (page 41) and thus likely to make a less than normal contribution to $U^\circ_{mic} - \mu^\circ_W$. The total contribution from the $n_C - 1$ CH_2 groups of the alkyl chain is thus likely to be only $n_C - 3$ to $n_C - 5$ times $-$ 800 cal/mole, leading to a predicted value for $U^\circ_{mic} - \mu^\circ_W$ of $+$(400 to 2000) $-$ 800 n_C cal/mole. Thus the positive constant term in equations 7-8 to 7-11 is readily accounted for, simply on the basis that n_C is larger than the number of carbon atoms of the alkyl chain that actually enter the micelle core.

Since W°_{mic} is the size-limiting factor in micelle formation, its dependence on micelle size has to be important. As will be shown in Chapter 9, micelle head groups are forced closer together by an increase in micelle size, at least for relatively small micelles. Since the average micelle size $\bar{m}$ increases with the length of the alkyl chain in a homologous series, head group separation also decreases with increasing n_C and, because there has to be repulsion between head groups, this means that dW°_{mic}/dn_C will be positive, qualitatively accounting for the increment of about -700 cal/mole per CH_2 group in equations 7-8 to 7-11 as compared to the increment of -800 cal/mole expected on the basis of $U^\circ_{mic} - \mu^\circ_W$ alone. A quantitative calculation would require an intimate knowledge of the forces between head groups, which is not generally available.

For ionic micelles a crude estimate of W°_{mic} can be made by assuming that electrostatic repulsion constitutes the entire repulsive force. To estimate W°_{mic} on this basis we shall assume that small ionic micelles can be considered to be spherical (which will be shown later to be not quite correct). The electrostatic charge will be equated with the average number of amphiphile ions per micelle, $\bar{m}$, and the repulsive free energy will be equated with the work of charging, as calculated by the Debye–Hückel theory (Tanford, 1961, chapter 7). This quantity, divided by $\bar{m}$, gives the value of W°_{mic} per mole of amphiphile, so that

$$W^{\circ}_{\text{mic}} = \frac{N\epsilon^2 \bar{m}}{2D}\left(\frac{1}{r} - \frac{\kappa}{1 + \kappa a}\right) \tag{7-13}$$

where r is the micelle radius for a micelle of average size, a is the radial distance of closest approach of counterions, κ is the Debye–Hückel reciprocal length (proportional to the square root of the ionic strength), ϵ is the unit of electronic charge, D is the dielectric constant of the solvent (in this case water), and N is Avogadro's number. The radius of the assumed spherical micelles required in the equation can be obtained from the values of $\bar{m}$ and estimates of the molar volume of the amphiphile molecules, with a liberal allowance for solvent incorporation between head groups near the micelle surface. The distance a is set equal to $r + 2.5$ Å, and the only experimental data required for the calculation are then the values of $\bar{m}$ as a function of n_C.

It should be emphasized that this is a crude method for estimating W°_{mic}, because of the high charge density at the micelle surface. Better estimates can be made by means of computer calculations, as discussed for instance by Stigter and Overbeek (1957) and Emerson and Holtzer (1967a). These estimates generally give lower values for W°_{mic} than equation 7-13. Lower values would also result from the binding of counterions to surface charges, which would make the effective micellar charge less than $\bar{m}$. The calculations using equation 7-13 should accordingly be considered as representing an upper limit to W°_{mic}, and we shall see that a more self-consistent interpretation of experimental data is obtained if one assumes that the actual value of W°_{mic} is about one-half the value given by equation 7-13. This is not an unreasonable result.[2]

It is probable that this calculation can also be applied to the N-alkyl betaines, for which the most reliable values of $\bar{m}$ are available over a con-

[2] Computer calculations have not been used here because they represent a refinement in the calculation without a change in the underlying model, which is a very artificial one because it treats the micelle surface as a uniformly charged layer, whereas it actually is an assembly of discrete charges that are quite far apart. A model exists for calculations based on a discrete charge distribution, but it contains unknown parameters that in our experience make the uncertainty of the calculation as large as that of equation 7-13.

If appropriate assumptions are made, a discrete charge model can be applied to non-spherical micelles. We have made some alternate calculations of W°_{mic} using the ellipsoidal model for ionic micelles, and the associated charge separations, that will be presented in Chapter 9 (Fig. 9-1). Within the limit of accuracy that can be claimed for any of these calculations, the estimates of W°_{mic} are similar to those obtained by use of equation 7-13.

siderable range of n_C. As can be seen from Fig. 7-2, the cmc of alkyl trimethyl ammonium bromide in the presence of 0.5*M* NaBr (only two experimental points are available) is virtually identical to the cmc of N-alkyl betaines with alkyl groups of the same chain length. The values of $\bar{m}$ are also very similar. This suggests that the repulsive force between the betaine head groups in water is similar in magnitude to the repulsive force between the ionic head groups of $RN(CH_3)_3^+$ at high ionic strength, and this is not unreasonable since the $—N(CH_3)_2^+—$ group of the betaines must lie at the very surface of the hydrophobic core of the micelles, whereas the terminal COO^- group can extend into the surrounding solvent with some freedom of motion, that is, it will resemble the diffuse layer of Br^- counterions surrounding $RN(CH_3)_3^+$ micelles at high ionic strength.

The use of equation 7-13 for W°_{mic} neglects specific ionic effects, and it is then evident that all parameters of micelle formation, including the cmc, $\bar{m}$, and W°_{mic} for any homologous series of amphiphiles at a given temperature, must be uniquely determined by two variables, the number of carbon atoms in the hydrocarbon chain (n_C) and the ionic strength I. The general expression for dW°_{mic}/dn_C thus may be written as

$$\frac{dW^\circ_{mic}}{dn_C} = \left(\frac{\partial W^\circ_{mic}}{\partial n_C}\right)_I + \left(\frac{\partial W^\circ_{mic}}{\partial \ln I}\right)_{n_C} \frac{d \ln I}{dn_C} \qquad (7\text{-}14)$$

For ionic micelles at high ionic strength, or for the N-alkyl betaines that we shall treat as such, the second term vanishes. Using the $\bar{m}$ values of Swarbrick and Daruwala (1969) to obtain $d\bar{m}/dn_C$ and appropriate dimensional parameters to evaluate dr/dn_C from it, dW°_{mic}/dn_C for the N-alkyl betaines is obtained directly from equation 7-13. Taking the mobile COO^- groups as equivalent to an ionic strength of 0.5, one obtains $dW^\circ_{mic}/dn_C \simeq 140$ cal/mole per CH_2 group. If this is coupled with the figure of about -800 cal/mole that is expected for $d(U^\circ_{mic} - \mu^\circ_W)/dn_C$, the calculated value for the n_C-dependent term of equation 7-8 becomes -660 cal/mole. As was indicated above, better agreement with experiment is obtained if W°_{mic} is taken to have about one-half the value given by equation 7-13, which would lead to a calculated increment in $\mu^\circ_{mic} - \mu^\circ_W$ of -730 cal/mole per CH_2 group.

To make a comparable calculation for the cmc values obtained in the absence of added salt, we must first note that the ionic strength is very low in these experiments, and the electrostatic repulsion between head groups is consequently much stronger. This leads to generally more positive values for $\mu^\circ_{mic} - \mu^\circ_W$ and larger values for the cmc, but is actually not the major factor that leads to the reduced slope in line F of Fig. 7-2, as compared to the other data in the figure. The reduced slope arises from the fact that the ionic strength in these experiments is determined by the concentration of free amphiphile, that is, by the cmc itself. Each individual determination in a homologous series thus represents a *different* ionic strength, and the last term in equation 7-14 becomes an important factor. To evaluate this factor we subdivide $\partial W^\circ_{\text{mic}}/\partial \ln I$ into a direct effect of ionic strength at constant $\bar{m}$, and an indirect effect resulting from the dependence of $\bar{m}$ on ionic strength, obtaining (from equation 7-13)

$$\left(\frac{\partial W^\circ_{\text{mic}}}{\partial \ln I}\right)_{n_C} = -\frac{N\epsilon^2\bar{m}}{4D}\frac{\kappa}{(1+\kappa a)} + \left(\frac{\partial W^\circ_{\text{mic}}}{\partial \bar{m}}\right)_{n_C}\left(\frac{\partial \bar{m}}{\partial \ln I}\right)_{n_C} \tag{7-15}$$

(In the interest of brevity, the explicit expression for $\partial W^\circ_{\text{mic}}/\partial \bar{m}$ has not been written.) Accurate values for $\bar{m}$ as a function of ionic strength are available for dodecyl sulfate (Emerson and Holtzer, 1965, 1967a), and equation 7-15 can thus be evaluated at $n_C = 12$. One obtains $\partial W^\circ_{\text{mic}}/\partial \ln I = -675$ cal/mole.

Combining equations 7-12 and 7-7 (which is a sufficient approximation here) and neglecting f_W as before,

$$RT\frac{d\ln \text{cmc}}{dn_C} = \frac{d(U^\circ_{\text{mic}} - \mu^\circ_W)}{dn_C} + \frac{dW^\circ_{\text{mic}}}{dn_C} \tag{7-16}$$

With equation 7-14 and $d \ln I/dn_C = d \ln \text{cmc}/dn_C$ this becomes

$$RT\frac{d\ln \text{cmc}}{dn_C} = \frac{[d(U^\circ_{\text{mic}} - \mu^\circ_W)]/dn_C + (\partial W^\circ_{\text{mic}}/\partial n_C)_I}{1-(RT)^{-1}(\partial W^\circ_{\text{mic}}/\partial \ln I)_{n_C}} \tag{7-17}$$

With the numerical value of -675 cal/mole for $\partial W^\circ_{\text{mic}}/\partial \ln I$ given above, the denominator of equation 7-17 becomes equal to 2.15, that is, equation 7-17 predicts that $d \ln \text{cmc}/dn_C$ for a homologous series in the absence of added salt should drop to slightly less than half the value expected from experiments at constant ionic strength. As before, we would get better agreement by assuming that W°_{mic} as calculated by equation 7-13 is too large by a factor of 2. The denominator of equation 7-17 then becomes 1.57 and the predicted value of $d \ln \text{cmc}/dn_C$ in the absence of added salt would be 64% of the value observed at constant ionic strength: this agrees almost exactly with the experimental decrease seen in Fig. 7-2.

These are not quantitative calculations, but they do serve to show that the observed dependence of $\mu^\circ_{\text{mic}} - \mu^\circ_W$ on n_C, at least for ionic or zwitterionic micelles, is not inconsistent with data for the simpler transfer reactions cited earlier in this book. They support the conclusion based on the results of Chapter 6, that the interior of small micelles has the characteristics of liquid hydrocarbon.

EFFECT OF TEMPERATURE

The effect of temperature on the cmc of ionic or zwitterionic micelles is small: $\bar{H}^\circ_{\text{mic}} - \bar{H}^\circ_W$ is close to zero at 25°C for all micelles that have been investigated (e.g., Goddard and Benson, 1957; Emerson and Holtzer, 1967b; Swarbrick and Daruwala, 1969; Shinoda et al., 1963). For nonionic micelles the enthalpy of micellization is positive (e.g., Corkill et al., 1964 a, b; Corkill et al., 1966; Muller and Platko, 1971). In all cases therefore the driving force for micelle formation is a positive entropy change, as is to be expected for a phenomenon that is a manifestation of the hydrophobic effect. The enthalpy change is temperature-dependent in all cases: values of $(\bar{C}^\circ_p)_{\text{mic}} - (\bar{C}^\circ_p)_W$ between -50 and -100 cal/deg-mole have been reported for all amphiphiles investigated.

The accuracy with which these thermochemical data can be determined is not as good as the accuracy of the free energy data described in the earlier part of this chapter. Moreover, there have been no systematic studies of homologous series of amphiphiles over a wide range of hydrocarbon chain lengths. An analysis of the results for the purpose of separating contributions arising from the hydrocarbon chain from those arising from head group interactions thus becomes impossible. For ionic micelles $\bar{S}^\circ_{\text{mic}} - \bar{S}^\circ_W$ is certainly smaller than the expected value of $\bar{S}^\circ_{\text{org}} - \bar{S}^\circ_W$ for the hydrocarbon chain (based on the data of Chapter 4) and the negative value of $(\bar{C}^\circ_p)_{\text{mic}} - (\bar{C}^\circ_p)_W$ is also smaller than the change in heat capacity that accompanies transfer of a hydrocarbon chain from water to an organic solvent. This means that the head group interaction is associated with a negative entropy change and a positive ΔC_p. This is not surprising, since the association of like charges increases the electrostatic potential and this should result in an increase in hydration and changes in entropy and heat capacity in the same direction as are associated with ion hydration in general (page 21). However, as indicated above, it is not possible to make a quantitative assignment of a given fraction of the observed results to this effect, and we cannot therefore make a meaningful estimate of the contribution to ΔS and ΔC_p that is associated with the hydrophobic effect.

At the present time, therefore, the available data on the effect of temperature on the free energy of micelle formation do not make a significant contribution to our knowledge of the process.

REFERENCES

Anacker, E. W. (1970). In *Cationic Surfactants*, E. Jungermann, Ed., Marcel Dekker, Inc., New York.

Balmbra, R. R., J. S. Clunie, J. M. Corkill, and J. F. Goodman. (1964). *Trans. Faraday Soc.*, **60**, 979.

Becher, P. (1967). In *Nonionic Surfactants*, M. J. Schick, Ed., Marcel Dekker, Inc., New York.

Corkill, J. M., J. F. Goodman, and S. P. Harrold. (1964a). *Trans. Faraday Soc.*, **60,** 202.

Corkill, J. M., J. F. Goodman, and J. R. Tate. (1964b). *Trans. Faraday Soc.*, **60,** 996.

Corkill, J. M., J. F. Goodman, R. Robson, and J. R. Tate. (1966). *Trans. Faraday Soc.*, **62,** 987.

Debye, P. (1949). *Ann. N.Y. Acad. Sci.*, **51,** 575.

Emerson, M. F., and A. Holtzer. (1965). *J. Phys. Chem.*, **69,** 3718.

Emerson, M. F., and A. Holtzer. (1967a). *J. Phys. Chem.*, **71,** 1898.

Emerson, M. F., and A. Holtzer. (1967b). *J. Phys. Chem.*, **71,** 3320.

Evans, H. C. (1956). *J. Chem. Soc.*, p. 579.

Geer, R. D., E. H. Eylar, and E. W. Anacker. (1971). *J. Phys. Chem.*, **75,** 369.

Goddard, E. D., and G. C. Benson. (1957). *Can. J. Chem.*, **35,** 986.

Hall, D. G., and B. A. Pethica. (1967). In *Nonionic Surfactants*, M. J. Schick, Ed., Marcel Dekker, Inc., New York.

Hill, T. L. (1964). *Thermodynamics of Small Systems*, Vol. 2. W. A. Benjamin, Inc., New York.

Klevens, H. B. (1953). *J. Am. Oil Chem. Soc.*, **30,** 74.

Mukerjee, P. (1967). *Adv. Colloid and Interface Sci.*, **1,** 241.

Muller, N., and F. E. Platko. (1971). *J. Phys. Chem.*, **75,** 547.

Shinoda, K., T. Nakagawa, B. Tamamushi, and T. Isemura. (1963). *Colloidal Surfactant*, Academic Press, New York.

Stigter, D., and J. T. G. Overbeek. (1957). *Proc. 2nd International Congress of Surface Activity*, Vol. I, p. 311, Butterworth and Co., London.

Swarbrick, J. and J. Daruwala. (1969). *J. Phys. Chem.*, **73,** 2627.

Swarbrick, J., and J. Daruwala. (1970). *J. Phys. Chem.*, **74,** 1293.

Tanford, C. (1961). *Physical Chemistry of Macromolecules*. John Wiley and Sons, New York.

EIGHT

MICELLE SIZE AND SIZE DISTRIBUTION

Molecular weights of micelles are readily measured by the same methods that are normally used for true macromolecules. The majority of existing data has been obtained by the use of light scattering, and almost all data are for micelles formed by amphiphiles with a single hydrocarbon chain. Only this type of system will be discussed in this chapter: it will be seen in Chapters 9 and 12 that amphiphiles with more than one alkyl chain per head group fall into a separate category with respect to micelle size and shape.

It appears at first sight that micelle molecular weights are irregular and unpredictable functions of the properties of the amphiphile molecule, contrasting sharply with the regularity that emerges from cmc measurements. Table 8-1, for example, shows the weight average aggregation number of a series of substituted ammonium salts, all with a decyl hydrocarbon chain, and all having comparable cmc values. The values of $\bar{m}_w$ are evidently greatly influenced by the substitution of CH_3 groups for H atoms on the ionic head group and by the change of counterion from Cl^- to Br^-. Another example is provided by Fig. 8-1, which shows representative data for the dependence of $\bar{m}_w$ on the length of the hydrocarbon chain. Micelle size is seen to increase with chain length in every case, but the magnitude of the effect differs greatly for the three systems. As was shown in Fig. 7-2, the dependence of the cmc on chain length in the same systems is virtually identical. Figures 8-2 and 8-3 show that similar unpredictable behavior extends to the effect of external variables, such as ionic strength and temperature, on the micelle aggregation number.

Perhaps the most important observation concerns the effect of total amphiphile concentration. In micellar systems in which the weight-average

Table 8-1. Micelle Aggregation Numbers for Decyl Ammonium Salts[a]

	ln cmc (mole fraction units)	$\bar{m}_w$
$C_{10}H_{21}NH_3^+Cl^-$	−7.40	78
$C_{10}H_{21}NH_3^+Br^-$	−8.15	1100
$C_{10}H_{21}NH_2(CH_3)^+Br^-$	−7.93	670[b]
$C_{10}H_{21}NH(CH_3)_2^+Br^-$	−8.44	69
$C_{10}H_{21}N(CH_3)_3^+Br^-$	−8.03	48

[a] From Geer et al. (1971). For decylammonium chloride the solvent was 0.1 *m* NaCl, 30°C, for all other data it was 0.5 *m* NaBr, 25°C. Molecular weights were determined by light scattering.
[b] Light-scattering plots were curved and a value of $\bar{m}_w = 270$ and a somewhat smaller cmc would have been obtained from data at lower amphiphile concentration.

degree of aggregation ($\bar{m}_w$) does not reach very large values one obtains light-scattering plots similar to those obtained for simple macromolecules, and they can be described in terms of a single value of $\bar{m}_w$, independent of concentration. In those systems, however, in which $\bar{m}_w$ rises to large values, anomalous light-scattering plots are often obtained, indicating a marked dependence of $\bar{m}_w$ on amphiphile concentration. An example is provided by Fig. 8-4, which shows a Debye light-scattering plot for hexadecyl heptaoxyethylene glycol monoether (Attwood, 1968). The pronounced curvature in

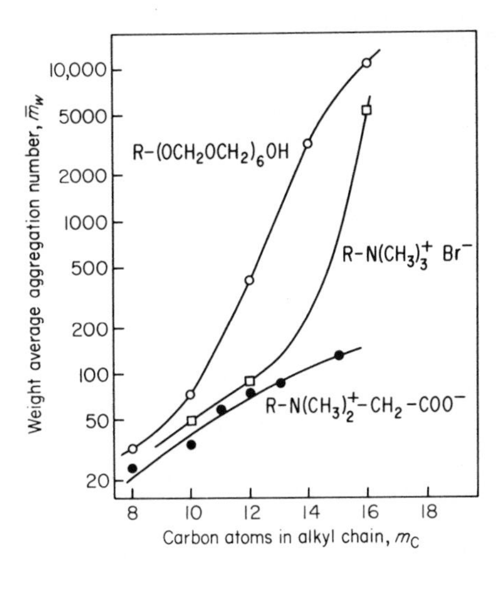

Fig. 8-1. Weight average micelle aggregation numbers, determined by light scattering, as a function of hydrocarbon chain length. The data for the alkyl hexaoxyethylene glycol monoethers are from Balmbra et al. (1964) and those for *N*-alkyl betaines are from Swarbrick and Daruwala (1970). The solvent in both cases was H_2O. The data for the alkyl trimethyl ammonium salts were obtained in 0.23 to 0.50 *M* NaBr, and are taken from Geer et al. (1971), Emerson and Holtzer (1967), and Anacker (1970). All data refer to 25°C.

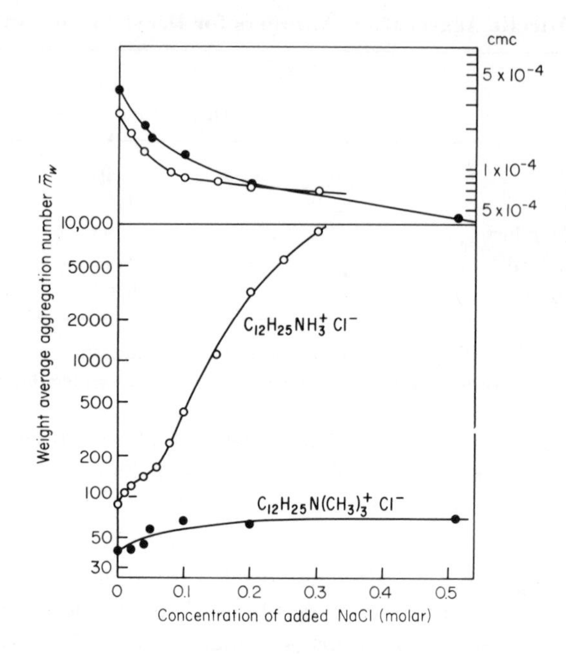

Fig. 8-2. Effect of added electrolyte on micelle aggregation numbers of two structurally similar ionic amphiphiles. The corresponding cmc values are shown at the top of the figure. Data for the ammonium salt are for 30°C (Kushner et al., 1957), and those for the trimethyl ammonium salt are for 23 to 25°C (Emerson and Holtzer, 1967; Kushner et al., 1957).

this plot indicates a huge increase in molecular weight with increasing concentration, leveling off at an $\overline{m}_w$ values of about 2600 at a concentration of 0.25 g/100 ml and above. Extrapolation to zero micelle concentration, however, leads to an $\overline{m}_w$ of only 145 or less. Most systems have not been as carefully investigated as this one, but similar effects have been observed, for instance, for the cationic detergents that can attain high $\overline{m}$ values (Geer et al., 1971). It should perhaps be noted that actual molecular weights in these systems are subject to some uncertainty because the second virial coefficient, representing the effect of thermodynamic nonideality as the concentration increases, cannot be determined from the experimental data when $\overline{m}_w$ changes with concentration. The possibility of making corrections by means of calculated second virial coefficients has been discussed by Mukerjee (1972).

Closer examination of these and other data indicate that the irregularities

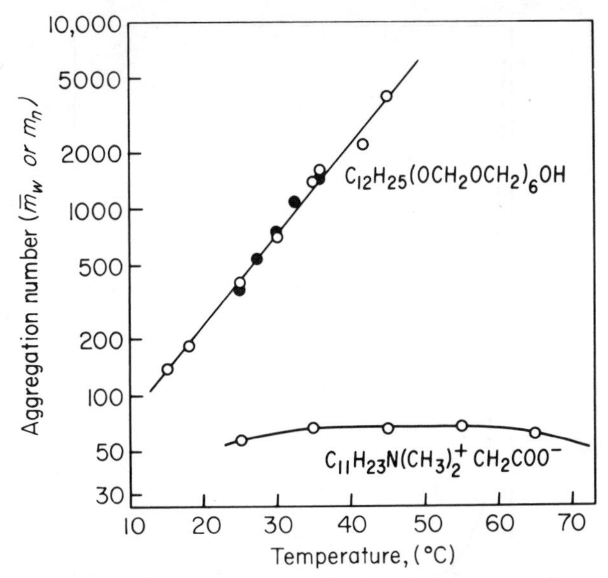

Fig. 8-3. Micelle aggregation number as a function of temperature. Data for *N*-undecyl betaine from Swarbrick and Daruwala (1970); data for dodecyl hexaoxyethylene glycol monoether from Balmbra et al. (1962) and Attwood et al. (1970). Open circles represent $\overline{m}_w$ by light scattering, filled circles $\overline{m}_n$ by osmotic pressure.

seen in the results we have cited are misleading and that micelle size in fact may follow definite rules that can be qualitatively defined as follows.

1. There are two distinct kinds of micelles: small micelles with $\overline{m}_w \sim 100$, and large micelles with $\overline{m}_w \gtrsim 1000$. All amphiphiles with single hydrocarbon chains can form both types. In many of the systems where no great increase in size is observed under dilute solution conditions, a transition to much larger micelles has been observed at very high amphiphile concentration by X-ray scattering (Reiss–Husson and Luzzati, 1964, 1966).

2. Factors that favor an increase in size produce relatively small increases in size as long as the system is within the realm of small micelles, but dramatic changes are observed as the upper limit for small micelles is approached and transition to large micelles occurs. (For example, note the change in slope at ionic strength 0.06 in the curve for dodecyl ammonium chloride in Fig. 8-2.)

3. The magnitude of the repulsive force between amphiphile head groups profoundly influences micelle size and the conditions required for transition from small to large micelles, in accordance with the principle of opposing

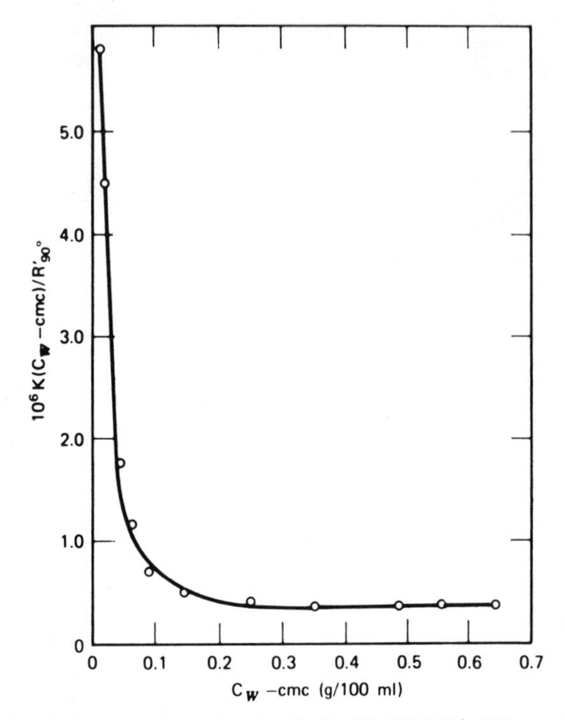

Fig. 8-4. Debye light-scattering plot for $C_{16}H_{33}(OCH_2CH_2)_7OH$ at 25°, taken from Attwood (1968). Extrapolation of the points at highest concentration to zero micelle concentration leads to $\overline{m}_w$ = 2600. Extrapolation of the steeply rising portion of the curve leads to $\overline{m}_w$ = 145.

forces (page 43), which identifies the repulsive force as the size-limiting factor. Most of the differences seen in Figs. 8-1 to 8-3 can be qualitatively explained in this way. For example, an increase in ionic strength decreases the repulsion between ionic head groups. It does so more effectively when the head group is $—NH_3^+$ than when it is $—N(CH_3)_3^+$ because counterions can approach more closely to the site of the cation charge of the $—NH_3^+$ group.

4. Other things being equal, an increase in the hydrocarbon chain length increases micelle size. This is again in accordance with the principle of opposing forces: an increase in the attractive force is accompanied by an increase in the repulsive force required to stop micelle growth. An example is provided by Fig. 8-1: whereas neither the C_{10}, nor the C_{12} derivative in the alkyltrimethylammonium bromide series undergoes transition to large

micelles, even at concentrations of several percent by weight, such a transition is observed for the C_{16} derivative.

5. An increase in total amphiphile concentration always favors formation of larger micelles, simply by virtue of the law of mass action. (A rigorous relation is provided by equation 8-9 below.) Whether the increase is dramatic, as in Fig. 8-4, or so small as to be virtually undetectable, is determined by the second principle given above.

MICELLE SIZE DISTRIBUTION

The continuous variability in $\bar{m}$ with external variables (especially the variation with amphiphile concentration) shows that micelles are not stoichiometric compounds, but aggregates capable of existing over a wide range of micelle sizes. It implies also that micelles existing in equilibrium with a monomeric amphiphile under a given set of conditions must be heterogeneous with respect to micelle size. The general result mentioned earlier, that only small changes in size occur when changes are occurring entirely within the realm of small micelles, coupled with the observation from the data of Fig. 8-4 that, at least for the particular amphiphile to which these data refer, the molecular weight again levels off when the transition to large micelles is complete, suggests that small and large micelles may be discrete populations and that micelles of intermediate size have relatively low stability.

Comparison of measurements by light scattering and osmotic pressure, in the few instances where it has been done (as in Fig. 8-3), has indicated that weight- and number-average molecule weights are about equal, which means that the size distribution cannot be unduly broad. However, this result may be misleading, because the data were probably obtained under conditions where most of the micelles present represent only one kind, that is, only the large micelles in the example of Fig. 8-3. Equations 8-5 to 8-7 below in fact indicate that different molecular weight averages could not have identical values if measurements are made in the region of transition from small to large micelles.

The distribution function for micelle size can in principle be rigorously related to the concentration dependence of molecular weight averages. We shall use A_m to designate a micelle of aggregation number m and $[A_m]$ for the corresponding concentration in mole fraction units. In terms of the parameters of Chapter 7,

$$X_{\mathrm{mic},m} = m[A_m] \tag{8-1}$$

$$X_{\mathrm{mic}} = \Sigma m[A_m] \tag{8-2}$$

the summation in equation 8-2 and elsewhere below extending over all micelle sizes that contribute significantly to the total micelle population. Equation 7-4 may then be rewritten as

$$RT \ln [\mathrm{A}_m] = -m(\mu^{\circ}_{\mathrm{mic},m} - \mu^{\circ}_W) + mRT \ln X_W f_W \qquad (8\text{-}3)$$

It is important to keep in mind that this equation assumes that micelles behave ideally: the left-hand side would have to include an activity coefficient if the possibility of interactions between micelles were to be incorporated, and this would represent an intractable factor in some of the subsequent derivations.

Equation 8-3 represents the distribution function for micelle size and shows that the breadth of the distribution depends on the functional relation between $\mu^{\circ}_{\mathrm{mic},m}$ and m. However, since $\mu^{\circ}_{\mathrm{mic},m}$ is by definition independent of amphiphile concentration, derivatives of $[\mathrm{A}_m]$ with respect to concentration variables are relatively simple. For example, neglecting the derivative of f_W,

$$d[\mathrm{A}_m]/dX_W = m[\mathrm{A}_m]/X_W \qquad (8\text{-}4)$$

This equation can be combined with the definitions of the various molecular size averages

$$\overline{m}_n = \Sigma m[\mathrm{A}_m]/\Sigma[\mathrm{A}_m] \qquad (8\text{-}5)$$

$$\overline{m}_w = \Sigma m^2[\mathrm{A}_m]/\Sigma m[\mathrm{A}_m] \qquad (8\text{-}6)$$

$$\overline{m}_z = \Sigma m^3[\mathrm{A}_m]/\Sigma m^2[\mathrm{A}_m] \qquad (8\text{-}7)$$

and with equation 8-2, to yield numerous useful relations of the type

$$d\overline{m}_n/d \ln X_{\mathrm{mic}} = \overline{m}_n(\overline{m}_w - \overline{m}_n)/\overline{m}_w \qquad (8\text{-}8)$$

$$d\overline{m}_w/d \ln X_{\mathrm{mic}} = \overline{m}_z - \overline{m}_w \qquad (8\text{-}9)$$

Moreover, the total amphiphile concentration (X_0) is equal to

$$X_0 = X_W + X_{\mathrm{mic}} \qquad (8\text{-}10)$$

and, since X_W is virtually constant and equal to the cmc (in mole fraction units) under the conditions where $\overline{m}_n$ and $\overline{m}_w$ are measured, the derivative $d \ln X_{\mathrm{mic}}$ in these equations can probably be replaced without significant error by $dX_0/(X_0 - \mathrm{cmc})$.

These equations show that the difference between $\overline{m}_z$, and $\overline{m}_w$ and $\overline{m}_n$ can be experimentally determined from the derivatives of $\overline{m}_n$ and $\overline{m}_w$ with respect to total concentration. Higher averages such as $\overline{m}_{z+1} = \Sigma m^4[\mathrm{A}_m]/\Sigma m^3[\mathrm{A}_m]$ can similarly be obtained from the higher derivatives of molecular weight with respect to total concentration, and all such averages taken together can be used to generate a size distribution function (references are cited in Tanford, 1961, p. 265). In principle, therefore, distribution functions are experimentally accessible, but the experimental difficulties are formidable. Because measurements to high amphiphile concentrations are required, thermodynamic nonideality cannot be avoided. We have already noted that this makes it difficult to determine unambiguous $\bar{m}$ values. A second difficulty is that the equations given here become incorrect if an activity coefficient has to be introduced on the left-hand side of equation 8-3, or if the presence of micelles at high concentrations affects f_W. An initial effort to use the concentration dependence of $\bar{m}$ for analysis of qualitative aspects of the size distribution in some micellar systems has been made by Mukerjee (1972).

DERIVATION OF A PLAUSIBLE DISTRIBUTION FUNCTION FOR SMALL MICELLES

Partial differentiation of equation 8-3 with respect to m at constant X_W and f_W relates the distribution function in a solution of given composition to the dependence of $\mu^{\circ}_{\mathrm{mic},m}$ on m. If X_W is set equal to the cmc, the micelle concentration to which the derivative refers is small, and the neglect of the activity coefficient on the left-hand side of equation 8-3 is probably not important. We shall consider the application of the derivative to a system of small *ionic* micelles in which the formation of large micelles at higher concentrations has not been observed. Since *very* small micelles are also unstable by virtue of the cooperative nature of micelle formation, such a system must be characterized by an optimal value of m, which we shall call m^*, at which $[\mathrm{A}_m]$ attains a maximum value. The value of m^* is obtained by setting $(\partial \ln [\mathrm{A}_m]/\partial m)X_W = 0$, yielding

$$RT \ln [\mathrm{A}_{m*}] = m^{*2}(d\mu^{\circ}_{\mathrm{mic},m}/dm)_{m=m*} \tag{8-11}$$

In analogy with equation 7-12, $\mu^{\circ}_{\mathrm{mic},m}$ may be considered to be the sum of attractive and repulsive factors, so that

$$RT \ln [\mathrm{A}_{m*}] = m^{*2}(dU^{\circ}_{\mathrm{mic},m}/dm)_{m=m*} + m^{*2}(dW^{\circ}_{\mathrm{mic},m}/dm)_{m=m*} \tag{8-12}$$

Since $RT \ln [\mathrm{A}_{m*}]$ is necessarily negative, equation 8-11 leads to the important conclusion that $\mu^{\circ}_{\mathrm{mic},m}$ must be a decreasing function of m. Since $dW^{\circ}_{\mathrm{mic},m}/dm$ is positive (page 55) equation 8-12 shows that $U^{\circ}_{\mathrm{mic},m}$ must also be a decreasing function of m. This is a reasonable result in view of the restricted freedom of motion of hydrocarbon chains in a small micelle, reference to which has already been made several times before. An increase in micelle size should relieve the restrictions to some extent. As we shall see, only very small values of $dU^{\circ}_{\mathrm{mic},m}/dm$ are obtained from actual experimental data.

If $W^\circ_{\mathrm{mic},m}$ and its dependence on m are known, a distribution function for micelle size can be generated by means of an iterative procedure. We first assume that $m^* = \bar{m}_w$ and assume a value for the fraction of amphiphile molecules actually in micelles of optimal size, that is, a value for the ratio $X_{\mathrm{mic},m^*}/X_{\mathrm{mic}}$. The value for $[A_{m^*}]$ when $X_W = \mathrm{cmc}$ is then obtained from equation 8-10. Equation 8-12 then yields an experimental value for $dU^\circ_{\mathrm{mic},m}/dm$. It is then necessary to introduce a functional relationship for the dependence of $U^\circ_{\mathrm{mic},m}$ on m. To do so we have made the same assumption that was made earlier for the calculation of W°_{mic} that the micelles can be considered to be spherical for the purpose of this calculation. We have further assumed that $U^\circ_{\mathrm{mic},m} - \mu^\circ_W$ can be considered to be proportional to the fraction of the micelle core volume that lies further than some critical distance from the hydrophilic surface layer, the hydrocarbon lying within that distance being too restricted to benefit from the transfer from water to the micelle interior. The equation for $U^\circ_{\mathrm{mic},m} - \mu^\circ_W$ then takes the form

$$(U^\circ_{\mathrm{mic},m} - \mu^\circ_W) = A(1 - B/r_0)^3 \tag{8-13}$$

A representing the value of $U^\circ_{\mathrm{mic},m} - \mu^\circ_W$ for a micelle of infinite size, and r_0 representing the radius of the micellar core which, for a spherical model, is proportional to $m^{1/3}$. Its value at $m = m^*$ is calculable from the known density of liquid hydrocarbon. The values of A and B are now obtained from equation 8-12 and equation 8-3, applied to $m = m^*$, in both of which A and B are now the only unknown parameters. Equation 8-13 and $W^\circ_{\mathrm{mic},m}$ together provide an expression for $\mu^\circ_{\mathrm{mic},m} - \mu^\circ_W$ as a function of m, which gives the distribution function by equation 8-3. Knowledge of the distribution function permits calculation of the ratios $m^*/\bar{m}_w$ and $X^\circ_{\mathrm{mic},m^*}/X_{\mathrm{mic}}$, which are then used as the basis for a second iteration of the calculation. Subsequent iterations can be performed as necessary.

We have used the data of Emerson and Holtzer (1967) for $C_{12}H_{25}N(CH_3)_3^+$ Br^- at ionic strength 0.1, 25°C, to generate a distribution function by this procedure. The experimental value of $\bar{m}_w$ is 73 and ln cmc = − 9.57 in mole fraction units. As before, we have used the Debye–Hückel approximation as a basis for the calculation of the repulsive free energy term. Since, in Chapter 7, best agreement with experimental data on the dependence of the cmc on alkyl chain length and ionic strength was obtained by using half the value of W°_{mic} predicted by equation 7-13, we have used half the value here also, with m in place of $\bar{m}$. The distribution function obtained in this way is shown in Fig. 8-5. It is sufficiently narrow so that $m^* = \bar{m}_n = \bar{m}_w = \bar{m}_z$ within about ±2. The value of $d\mu^\circ_{\mathrm{mic},m}/dm$ at $m = m^*$ is only −1.9 cal/mole. With the calculated value of 8.0 cal/mole/mole for $dW^\circ_{\mathrm{mic},m}/dm$ this leads to $dU^\circ_{\mathrm{mic},m}/dm = -9.9$ cal/mole/mole. These are very small

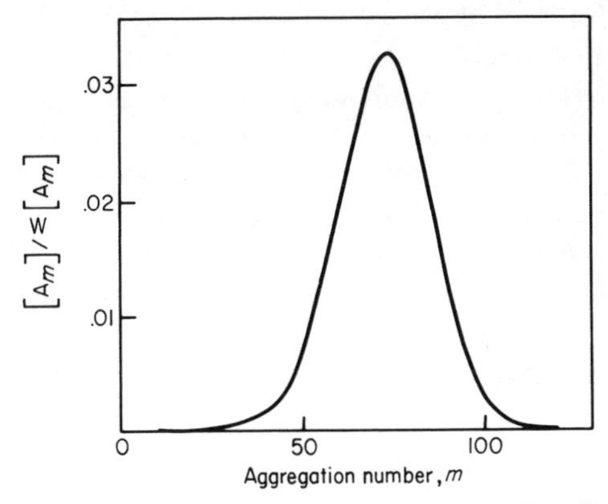

Fig. 8-5. Hypothetical micelle size distribution function for $C_{12}H_{25}N(CH_3)_3{}^+Br^-$ in 0.1 *M* NaBr. The distribution function is based on equation 7-13 for W°_{mic} and equation 8-13 for $U^\circ_{\text{mic}} - \mu^\circ_W$. The variable parameters in the latter equation were fixed by the fact that both $\mu^\circ_{\text{mic}} - \mu^\circ_W$ and $d\mu^\circ_{\text{mic}}/dm$ are known from experimental data.

figures indeed when compared to the value of $\mu^\circ_{\text{mic}} - \mu^\circ_W$ itself. Thus $\mu^\circ_{\text{mic},m*} - \mu^\circ_W$ is -5530 cal/mole. Note that this is only slightly more positive than the value of $\mu^\circ_{\text{mic}} - \mu^\circ_W$ obtained by equation 7-6, which is -5550 cal/mole.

The calculation made here is, of course, a crude one, and not to be taken seriously. However, Fig. 8-5 can be taken as an indication of how a distribution function for small micelles *might* look without doing violence to any known experimental facts.

GROWTH OF MICELLES TO LARGE SIZE

With a bellshaped distribution function such as that shown in Fig. 8-5 there can be no significant dependence of molecular size on concentration. A steep rise in $\bar{m}_w$ with concentration, as in Fig. 8-4, in fact requires, by equation 8-9, that $\bar{m}_z$ must be at least 50 to 100% higher than $\bar{m}_w$ in the transition region from small to large micelles. If Fig. 8-5 is representative of distribution functions for small micelles (which, of course, it may not be, especially for a nonionic amphiphile) this would mean that large micelles must represent a completely distinct population, as was already suggested earlier. The overall distribution function would be bimodal, consisting of a curve such as that of Fig. 8-5 together with a second curve (very likely, a much broader one), characterized by a much larger value of m^*.

It should be pointed out that the molecular weight measurements alone do not require that what we have called "large micelles" are necessarily true micelles with a single large hydrophobic core. Large molecular weights could alternatively be the result of aggregation of small micelles, to form clusters in which the intrinsic structure of the small micelles is unaltered. This possibility is, however, rather unlikely, in view of the fact that repulsive forces between amphiphile head groups are a prerequisite for micelle formation per se. To account for the transition from small micelles to large single micelles requires only that the repulsive force between head groups (at a given micelle size) be reduced in magnitude. To account for the formation of aggregates of small micelles, on the other hand, requires postulation of an attractive force between similar head groups located on separate small micelles. Shape measurements discussed in Chapter 9 also argue against the aggregation mechanism.

REFERENCES

Anacker, E. W. (1970). In *Cationic Surfactants*, E. Jungermann, Ed., Marcel Dekker, inc., New York.

Attwood, D. (1968). *J. Phys. Chem.*, **72,** 339.

Attwood, D., P. H. Elworthy, and S. B. Kayne. (1970). *J. Phys. Chem.*, **74,** 3529.

Balmbra, R. R., J. S. Clunie, J. M. Corkill, and J. F. Goodman. (1962). *Trans. Faraday Soc.*, **58,** 1661.

Balmbra, R. R., J. S. Clunie, J. M. Corkill, and J. F. Goodman. (1964). *Trans. Faraday Soc.*, **60,** 979.

Emerson, M. F., and A. Holtzer. (1967). *J. Phys. Chem.*, **71,** 1898.

Geer, R. D., E. H. Eylar, and E. W. Anacker. (1971). *J. Phys. Chem.*, **75,** 369.

Kushner, L. M., W. D. Hubbard, and R. A. Parker. (1957). *J. Res. Natl. Bur. Stand.*, **59,** 113.

Mukerjee, P. (1972). *J. Phys. Chem.*, **76,** 565.

Reiss-Husson, F., and V. Luzzati. (1964). *J. Phys. Chem.*, **88,** 3504.

Reiss-Husson, F., and V. Luzzati. (1966). *J. Colloid and Interface Sci.*, **21,** 534.

Swarbrick, J., and J. Daruwala. (1970). *J. Phys. Chem.*, **74,** 1293.

Tanford, C. (1961). *Physical Chemistry of Macromolecules.* John Wiley and Sons, New York.

NINE

MICELLE SHAPE

The frictional properties of small micelles resemble those of globular proteins. For example, several determinations of the intrinsic viscosity of dodecyl sulfate micelles in $0.1M - 0.2M$ NaCl at 25°C lead to the result that $[\eta] = 3.3 \pm 0.1$ cc/g (Parker and Wasik, 1958; Mukerjee, 1964; Tokiwa and Ohki, 1967). The frictional coefficient ratio, $f/f_{\min}$, under similar conditions, derived from sedimentation and diffusion data, is 1.16 (Tokiwa and Ohki, 1967). These results indicate that these micelles, like globular proteins, are compact, close to spherical, and that little solvent is incorporated in the hydrodynamic particle. (The procedure for interpretation of such data is described by Tanford, 1961, Chapter 6.)

The large micelles discussed in the previous chapter behave quite differently. For example, $C_{12}H_{25}NH_3^+Cl^-$, in $0.3M$ NaCl ($\bar{m}_w = 900$; see Fig. 8-2), has $[\eta] = 45$ cc/g (Kushner et al., 1957). At low ionic strength, on the other hand, where this amphiphile forms small micelles, $[\eta] = 3.5$ cc/g. For $C_{12}H_{25}N(CH_3)_3^+Cl^-$, which does not form large micelles as the ionic strength increases, $[\eta]$ does not rise to large values. Another example is provided by the nonionic oxyethylene glycol monoethers, $R\text{–}(OCH_2CH_2)_n\text{–}OH$. When R represents a long hydrocarbon chain, these substances form very large micelles (Fig. 8-1), and the micelle size increases greatly with temperature (Fig. 8-3). The intrinsic viscosity rises to correspondingly large values. Thus with $R = C_{16}H_{33}$ and $n = 7$, $[\eta] = 105$ cc/g at 25°C, and 176 and 260 cc/g, respectively, at 30 and 35°C (Attwood, 1968).

Hydrodynamic data of this kind are supported by all other physical measurements that reflect the extent of the domain of a dissolved particle. For example, the dramatic effect of total amphiphile concentration on the micelle molecular weight of $C_{16}H_{33}(OCH_2CH_2)_7OH$, illustrated by Fig. 8-4, is accompanied by an equally dramatic change in light scattering dissymmetry (Attwood, 1968), indicating a transition from compact to much more extended micelles.

There are in principle three ways of accounting for these large $[\eta]$ values. (1) The large micelles may, like small micelles, be sparingly solvated, but highly asymmetric in shape, for example, they may be long rodshaped particles. (2) The large micelles may be symmetrical but highly solvated, resembling random coils of polymer molecules. (3) As suggested at the end of the previous chapter, large micelles may arise from self-association of small globular micelles, instead of representing formation of a single particle with a homogeneous hydrophobic core. Only the first of these interpretations is reasonable. Formation of something resembling a solvent-penetrated random coil does not seem possible here since the basis for micelle formation rests on the necessity of avoiding contact between the hydrophobic portions of amphiphile molecules and the solvent, with formation of solventfree domains resembling liquid hydrocarbon droplets (Chapter 6). Self-aggregation is not excluded by the physical measurements, provided that the association is predominantly linear, leading to a "string of beads" type of product. This would seem to be an inherently unlikely process: the compactness of the small micelles and the uniformity of their surfaces would suggest that self-association, if it occurs, should lead to irregular clusters. Such clusters would not be expected to have intrinsic viscosities more than two or three times as large as the intrinsic viscosity of the compact micelles from which they are formed. We thus conclude that large soluble micelles (formed by amphiphiles with a single hydrocarbon chain) are probably quite asymmetric particles, but consisting, like smaller micelles, of a singly connected core with a continuous surface. The most likely shape is one resembling a long thin rod, as was first proposed on the basis of light-scattering measurements by Debye and Anacker (1951). (See also the review by Anacker, 1970.)

Stigter (1966) has made a detailed analysis of the intrinsic viscosity of dodecyl ammonium chloride micelles as a function of the changes in $\overline{m}_w$ that accompany changes in ionic strength (Fig. 8-2). His conclusion is that the micelles are rodlike, but that the rods must possess considerable flexibility. He points out that such flexibility would increase the entropy of the micelles, without significantly altering local interactions. Flexibility is perhaps also indicated by the analysis of the angular dependence of light scattering from cetyltrimethylammonium bromide micelles at high ionic strength (Anacker, 1970, p. 241). The data fit the theoretical scattering curve for a stiff rod, but the molecular weight calculated from the dimensions of the rod is less than the experimental molecular weight, corrected for nonideality on the basis of micelle shape.

Amphiphiles with Two Hydrocarbon Tails

All of the preceding results, and all the data of Chapter 8, refer to amphiphile molecules or ions with a single hydrocarbon tail. The limited data available for amphiphile molecules with two hydrocarbon tails per head group indicate quite different behavior. The substances in this category that have been studied are phospholipids of biological origin and, because of their special interest for the subject of this book, will be considered separately in Chapter 12. It may be noted here, however, that amphiphiles of this kind tend to form large planar bilayers, with hydrophilic head groups on the two external surfaces, and a layer of hydrocarbon between them. The bilayers can be folded to form essentially spherical vesicles, containing a solvent-filled cavity. These vesicles can have a very large size without becoming unduly asymmetric. One preparation of such vesicles, formed from diacyl phosphatidyl choline (heterogeneous with respect to the hydrocarbon chains) has been studied by Huang (1969) and found to have $\bar{m} \simeq 3000$, but an intrinsic viscosity of only 4.1 cc/g.

X-RAY DIFFRACTION. ORDERED STRUCTURES AT HIGH AMPHIPHILE CONCENTRATION

X-ray diffraction is the principal tool available to the physical chemist for the investigation of ordered structures, but limited information about large particles in solution can also be obtained from measurements at small scattering angles. The method has been applied to amphiphile micelles in aqueous solution by Reiss–Husson and Luzzati (1964, 1966). The major contribution of this work has been to show that the transition from small globular micelles (spheres and ellipsoids could not be distinguished) to rodlike micelles with increasing amphiphile concentration is a general phenomenon for most ionic amphiphiles, even in the absence of added salt: the transition occurs under these conditions at much higher concentrations than those used to detect similar transitions at high ionic strength by hydrodynamic measurements.

When the concentration of amphiphile is raised to sufficiently high values (typically 20 to 40% by weight for amphiphiles with a single hydrocarbon chain), ordered liquid crystalline phases are formed. Luzzati and co-workers have studied a number of such systems, and a summary of their results has been presented by Luzzati (1968). The first such phase formed generally represents an ordering of the rodlike micelles already present in solution: the rods are roughly parallel and arranged hexagonally in the plane perpendicular to their length. As the amphiphile concentration is further increased (to 50% by weight and higher), several successive phase transitions occur. Their structures have not been determined unambiguously, but one of them may be an inverted type of hexagonal phase in which the *solvent* (*water*) occupies roughly cylindrical regions, the remainder of the space being filled by the amphiphile molecules with their head groups at the boundary of the solvent cylinders. As the water content is further depleted, it becomes impossible to maintain a relatively large separation between head groups, and a lamellar phase consisting of extended bilayers is formed. Under completely anhydrous conditions, various crystalline arrays of these bilayers are formed: most of them are stable at high temperatures only.

GEOMETRIC CONSIDERATIONS

The foregoing observations concerning the relation between micelle shape and size, and the effect of having two rather than one hydrocarbon chain attached to the hydrophilic head group, are greatly clarified on the basis of simple geometric considerations (Tartar, 1955; Tanford, 1972). The following assumptions underlie the procedure employed:

1. The micelle contains a hydrophobic core consisting entirely of portions of the hydrocarbon chains. It is assumed that the terminal CH_3 groups are always contained in this core, but one or more methylene groups near the amphiphile head group may not be. It is assumed that no solvent enters the core. The physical properties of the core resemble those of a droplet of liquid hydrocarbon, and this means that the volume of the core is calculable. For a micelle containing m' hydrocarbon chains the core volume V in cubic angstroms is

$$V = (27.4 + 26.9n_C')m' \qquad (9\text{-}1)$$

where n_C' represents the number of carbon atoms of the hydrocarbon chain that are embedded in the hydrophobic core which, as already noted, may be less than n_C, the total number of carbon atoms in the chain. The numerical parameters of equation 9-1 are based on volume measurements of Reiss–Husson and Luzzati (1964). Essentially the same relation would have been obtained on the basis of the published densities of liquid hydrocarbons or even by the empirical volume additivity rule of Traube (1899) (discussed by Cohn and Edsall, 1943, p. 147).

2. Since no hole can exist at the center of the micelle, one dimension is always limited by the maximum possible extension of a hydrocarbon chain. This distance is obtained from the distance of 2.53 Å between alternate carbon atoms of a fully extended chain, with the addition of the van der

Waals radius of the terminal methyl group (2.1 Å) and one-half the bond length to the first atom not contained within the hydrophobic core ($\simeq$0.6 Å). The maximum length l_{max} for a chain with n'_C embedded carbon atoms, in angstroms, becomes

$$l_{max} = 1.5 + 1.265n'_C \tag{9-2}$$

3. The surface area per amphiphile head group (S/m) is a critical parameter in the thermodynamics of micelle formation in that it is a measure of the separation between adjacent head groups. The principle of opposing forces (Chapter 7) will dictate an optimal value for this parameter. Repulsion between head groups will tend to increase S/m, but will become unimportant when S/m becomes sufficiently large. When S/m becomes large there will necessarily be contact between water molecules and the core surface, and a consequent pressure to reduce S/m. The optimal value of S/m will be determined by proper balance between these factors. An additional thermodynamic factor is the total amphiphile concentration: an increase in concentration represents a pressure for an increase in micelle size (equation 8-7) even though this may, and usually does, entail some decrease in S/m.

Globular Micelles

The radius of the hydrophobic core of a spherical micelle may not exceed l_{max} and the maximum number of hydrocarbon chains per micelle is thus uniquely determined for each value of n'_C by combination of equations 9-1 and 9-2. For amphiphiles with a single hydrocarbon chain this is equivalent to the maximum value for the micelle aggregation number, m. Table 9-1 shows calculated values for several values of n'_C and comparison of these with experimental micelle aggregation numbers immediately shows that most of the common small compact micelles cannot be truly spherical in shape. For example, the weight-average aggregation number ($\overline{m}_w$) of dodecyl sulfate micelles ranges from 62 in the absence of added salt to 126 in 0.5M NaCl (Mysels and Princen, 1959; Emerson and Holtzer, 1967). The maximum number for a spherical micelle depends on the choice of a value for n'_C, but even with $n'_C = n_C = 12$ the maximum value of m is only 56. With a more realistic choice, the maximal value is reduced, for example, with $n'_C = 10$, it is 40.

To incorporate a larger number of hydrocarbon chains in a micelle, a distortion in the micelle shape is required. The simplest possibilities are ellipsoids of revolution: the minor semiaxis b_0 of an ellipsoid may not exceed l_{max}, but the major semiaxis a_0, is not limited, and an increase in the available volume is thus provided. Table 9-1 shows pertinent calculations, for selected

Table 9-1. Maximal Micelle Aggregation Numbers for Spherical and Ellipsoidal Micelles[a, b]

n'_C =	6	10	12	15	20
Sphere,[c] $r_0 = l_{max}$	17	40	56	84	143
Ellipsoids, $b_0 = l_{max}$					
Prolate, $a_0/b_0 = 1.25$	21	50	70	105	178
$a_0/b_0 = 1.5$	25	60	84	126	214
$a_0/b_0 = 1.75$	29	70	97	146	250
$a_0/b_0 = 2.0$	33	80	111	167	285
Oblate, $a_0/b_0 = 1.25$	26	63	87	131	223
$a_0/b_0 = 1.5$	38	90	125	188	321
$a_0/b_0 = 1.75$	51	123	171	256	437
$a_0/b_0 = 2.0$	67	160	223	335	570

[a] n'_C represents the number of carbon atoms in that portion of the alkyl chain that is incorporated in the hydrophobic core. This will generally be less than the total length of the hydrocarbon chain.

[b] The tabulated figures are the number of hydrocarbon chains per micelle. This is equal to the micelle aggregation number m for amphiphiles with a single hydrocarbon chain. It would be equal to $2m$ for amphiphiles with two chains, each of n'_C carbon atoms.

[c] It is actually physically impossible to form a spherical micelle with $r_0 = l_{max}$ and maximal aggregation numbers for physically possible small micelles are thus even smaller than the tabulated figures.

values of n'_C, for both prolate and oblate ellipsoids. It is seen that only small values of the ellipsoidal axial ratio (a_0/b_0) are required for a substantial gain in the micelle aggregation number, sufficient to account for experimentally observed values.

Formation of an ellipsoid ($b_0 = l_{max}$) from a sphere ($r_0 = l_{max}$) is accompanied by a decrease in the surface area per hydrocarbon chain (S/m') and, as a consequence, the separation between head groups decreases. Since head groups extend away from the surface of the hydrophobic core, one is generally interested in knowing the value of S/m' at some distance d outside the core surface, that is, at the surface of a sphere with radius $r = r_0 + d$ or at the surface of an ellipsoid defined by semiaxes $a = a_0 + d$, $b = b_0 + d$. Figure 9-1 represents a typical example of the dependence of S/m' on m', as increasing m' requires formation of ellipsoids with increasing axial ratio. The calculations have been made with $n'_C = 12$ and $d = 2$ Å. The values of m' are those given in Table 9-1 for spheres and ellipsoids with r_0 or $b_0 = l_{max}$. Additional data have been generated by making calculations for globular

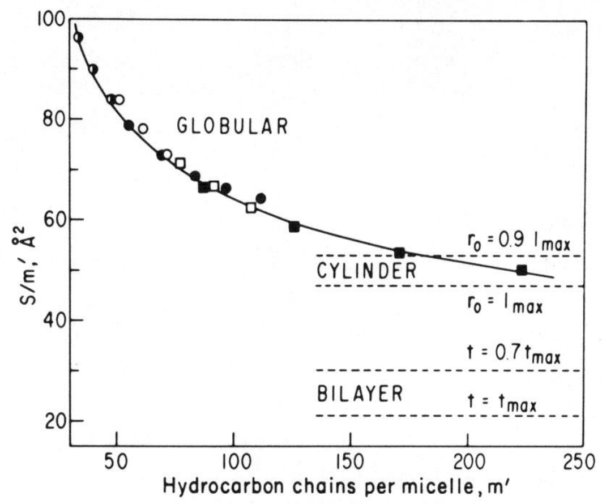

Fig. 9-1. Surface area per hydrocarbon chain as a function of micelle size and shape. Calculations are for $n_C' = 12$ and represent areas at a distance of 2 Å outside the hydrophobic core. The different symbols refer to calculations for spherical and ellipsoidal shapes, both with r_0 or b_0 equal to l_{max} (Table 9-1) and with these dimensions somewhat reduced. Surface areas for cylindrical micelles and bilayers are independent of aggregation number, apart from end effects, which have not been taken into consideration. In using these calculations it is important to distinguish between micelles formed by amphiphiles with a single hydrocarbon chain, for which the number of head groups is equal to the number of hydrocarbon chains, and amphiphiles with two hydrocarbon chains, for which the number of head groups is only $m'/2$, so that the area *per head group* is twice the area per hydrocarbon chain (taken from Tanford, 1972).

micelles with hydrophobic cores smaller than maximal size, that is, spheres with $r_0 < l_{max}$ and ellipsoids with $b_0 < l_{max}$. The figure also contains surface area calculations for cylinders and planar bilayers, which will be considered subsequently.

For amphiphiles with a single hydrocarbon chain Fig. 9-1 represents a plot of surface area per head group (S/m) versus aggregation number (m). When such amphiphiles form globular micelles, $\bar{m}$ is generally of order 100, and the figure shows that for micelles in this size range, S/m is close to being a single-valued function of m at a given value of n_C, only minimally affected by the actual micelle shape. It is evident that in this situation spherical and ellipsoidal micelles represent a family of conformations within which the micelle aggregation number is continuously variable, with a parallel variation in S/m. Any actual system that forms globular micelles will undoubtedly consist of a distribution of sizes and corresponding shapes about the optimal value of S/m, that is, a distribution such as is illustrated by Fig. 8-5, and will involve a distribution of shapes as well as aggregation number. The smallest micelles may be spherical, but micelles of average or larger size will be prolate or oblate ellipsoids, with asymmetry increasing as micelle size increases.

Cylindrical Micelles

Cylindrical micelles can grow in length without limit. Ignoring effects of the cylinder ends, S/m' is independent of length. The actual value of S/m' is smaller than for globular micelles, in the range of micelle size normally observed for such micelles, as is illustrated for $n'_C = 12$ in Fig. 9-1. On the other hand, as the micelle size increases, the surface area available for a globular micelle becomes about equal to that for a cylindrical micelle. When this occurs, a transition to cylindrical micelles is indicated and, because the latter can grow to much larger size without further decrease in specific surface area, a sharp increase in average size is predicted. In the transition zone itself a bimodal distribution, representing an equilibrium between globular and cylindrical micelles, might be expected. These predictions are consistent with the previously cited experimental observations for single-chain amphiphiles under conditions where the formation of large micelles occurs.

Bilayers and Vesicles

Bilayers can also accommodate a limitless number of amphiphile molecules without alteration in the available surface area per amphiphile. Values of S/m' are generally much smaller than for any other type of micelle, even if the thickness (t) of the hydrophobic core is made substantially less than its maximum value ($t_{max} = 2l_{max}$), as would be true if the hydrocarbon chains are not fully extended or not perpendicular to the bilayer surface. The values of S/m' are essentially independent of n'_C, that is, the values shown in Fig. 9-1 apply to any value of n'_C. For bilayers there is no gain in S/m' as one moves away from the surface of the hydrophobic core.

Bilayers are capable of forming spherical vesicles with an internal solvent-filled cavity. One bilayer surface is expanded in this process and the other is contracted. The net increase in S/m' is small. Even if the radius of the internal cavity is as small as the thickness of the bilayer core, the increase in S/m' is only 15%. As the overall vesicle size increases, S/m' approaches the value for an infinite planar bilayer.

Amphiphiles with Two Alkyl Chains

Figure 9-1 indicates that amphiphiles with a single hydrocarbon chain will form bilayers or vesicles only under extreme conditions, since an approximately twofold reduction in surface area is entailed. This is in agreement with the experimental observation that single-chain amphiphiles form lamellar structures only as one approaches anhydrous conditions. The situation is, however, quite different for amphiphiles with two alkyl chains per head group. For such substances the number of molecules per micelle is $m = m'/2$ and the surface area *per head group*, S/m, is therefore twice as large as the surface area *per hydrocarbon chain* given in Fig. 9-1. When two-chain amphiphiles are in a bilayer arrangement they have about the same surface area per head group as single-chain amphiphiles have in globular micelles.

There is no reason to believe that the optimal value of S/m for micelles formed by two-chain amphiphiles will be larger than the value of S/m for single chain amphiphiles with comparable head groups. On the contrary, a somewhat smaller value is to be expected when equal chain lengths are involved, because the value of $U^{\circ}_{mic} - \mu^{\circ}_{W}$ for a double-chain molecule is about 60% more negative than for a single chain amphiphile (Fig. 7-3), so that balance between the attractive and repulsive components of the chemical potential would be expected to occur at a somewhat larger value of W°_{mic}. The bilayer arrangement should thus be the optimal form for two-chain amphiphiles: globular or cylindrical micelles would have much too large a surface area.

This is undoubtedly the reason for the formation of bilayers by biological lipids. As noted above, micelles based on a bilayer arrangement can be expected to form closed vesicles, in order to leave no exposed hydrocarbon-water interface around an exposed edge. This phenomenon is indeed frequently observed, as will be seen in Chapter 12.

REFERENCES

Anacker, E. W. (1970). In *Cationic Surfactants*, E. Jungermann, Ed., Marcel Dekker, Inc., New York.

Attwood, D. (1968). *J. Phys. Chem.*, **72,** 339.

Cohn, E. J., and J. T. Edsall. (1943). *Proteins, Amino Acids and Peptides*. Reinhold, New York.

Debye, P., and E. W. Anacker. (1951). *J. Phys. and Colloid Chem.*, **55,** 644.

Emerson, M. F., and A. Holtzer. (1967). *J. Phys. Chem.*, **71,** 1898.

Huang, C. (1969). *Biochemistry*, **8,** 344.

Kushner, L. M., W. D. Hubbard, and R. A. Parker. (1957). *J. Res. Natl. Bur. Stand.*, **59,** 113.

Luzzati, V. (1968). In *Biological Membranes*, D. Chapman, Ed., Academic Press, New York, Chapter 3.

Mukerjee, P. (1964). *J. Colloid Sci.*, **19,** 722.

Mysels, K., and L. Princen. (1959). *J. Phys. Chem.*, **63,** 1696.

Parker, R. A., and S. P. Wasik. (1958). *J. Phys. Chem.*, **62,** 967.

Reiss-Husson, F., and V. Luzzati. (1964). *J. Phys. Chem.*, **68,** 3504.

Reiss-Husson, F., and V. Luzzati. (1966). *J. Colloid and Interface Sci.*, **21,** 534.

Stigter, D. (1966). *J. Phys. Chem.*, **70,** 1323.
Tanford, C. (1961). *Physical Chemistry of Macromolecules.* John Wiley and Sons, New York.
Tanford, C. (1972). *J. Phys. Chem.*, **76,** 3020.
Tartar, H. V. (1955). *J. Phys. Chem.*, **59,** 1195.
Tokiwa, F., and K. Ohki. (1967). *J. Phys. Chem.*, **71,** 1343.
Traube, J. (1899). *Samml. Chem. u. Chem. Tech. Vortr.*, **4,** 255.

TEN

MIXED MICELLES

Because the driving force for association between amphiphile molecules is nonspecific, and because the resultant micelle has a liquidlike core, micelles containing mixtures of amphiphile molecules should be formed readily. If we confine our attention to amphiphiles with linear hydrocarbon tails, there would seem to be no reason for any significant restriction on the nature of the head group in the formation of mixed micelles, other than the fundamental restriction that the interaction between head groups be repulsive. Thus one would not expect to observe formation of mixed soluble micelles when cationic and anionic amphiphiles are mixed in about equimolar amounts. Most other kinds of mixed micelles are in fact readily formed (Shinoda et al., 1963; Becher, 1967), and amphiphile molecules or ions with short hydrocarbon tails, which alone do not form sufficiently large micelles to be recognized as such, are readily incorporated in micelles formed from amphiphile molecules with long hydrocarbon tails.

For the purposes of this book we do not require a rigorous thermodynamic treatment of the formation of mixed micelles, as outlined by Hall and Pethica (1967). Quantitative predictions would in any event be impossible at the present time because the relation between the chemical potential $\mu_{i,\mathrm{mic}}$ of an amphiphile i in a mixed micelle and the corresponding chemical potential in a pure micelle of the same amphiphile depends on knowledge of the repulsive interactions between like and unlike head groups (i.e., W'_{mic} of equation 7-12), which we do not now possess.

The fundamental aspects of the formation of mixed micelles can be approached by making the approximation that the micelles constitute a separate phase, and that the free energy of the dispersal of this phase into small particles can be ignored. As has been shown, this is a good approximation in systems in which the micelle size is large. When this approximation is made, the thermodynamic treatment becomes analogous to the familiar treatment of liquid-vapor equilibria for liquid solutions, as described in any textbook of physical chemistry. Thus we can describe an ideal mixed micelle as one in which Raoult's law is obeyed: where $X_{i,\mathrm{mic}}$ is the mole fraction of amphiphilic component i in the mixed micelle and $X_{i,W}$ the mole fraction of the same component in monomeric state in the aqueous solution,

$$X_{i,W} = X_{i,\mathrm{mic}} X_{i,W}^{\circ} = X_{i,\mathrm{mic}} (\mathrm{cmc})_i^{\circ} \tag{10-1}$$

In this equation $X_{i,W}^{\circ}$ is the free concentration of amphiphile i in equilibrium with pure micelles which, as we have seen, is essentially the same as the critical micelle concentration for pure amphiphile i. Equation 10-1 shows that the concentration of free amphiphile in equilibrium with a mixed micelle will always be smaller than $(\mathrm{cmc})_i^{\circ}$.

Equation 10-1 allows us to construct phase diagrams analogous to those for liquid-vapor equilibria, relating the composition of mixed micelles to the composition of the solution in equilibrium with them. This has been done in Fig. 10-1 for mixtures of two polyoxyethylene-glycol derivatives differing only in the length of the hydrocarbon chain. The ordinate in the figure is the sum of the concentrations of the free amphiphiles in solution, in equilibrium with mixed micelles. This may be regarded as an effective cmc for the mixed micelle and, by equation 10-1, is necessarily a linear function of the mole fraction of either component in the micelle. The composition of the free amphiphile in solution is of course quite different from that of the micelle, being much richer in the component with the larger $(\mathrm{cmc})_i^{\circ}$. The situation is exactly analogous to the enrichment of the vapor phase by the more volatile component of a binary liquid mixture in equilibrium with it.

Experimental data for comparison with the predictions of Fig. 10-1 have not been reported. Experimental data for mixtures of anionic amphiphiles with identical head groups and different alkyl chains have however been obtained, and they agree quite well with predictions based on ideal mixing in the micelle (Shinoda et al., 1963). The experiments were conducted in the absence of added salt and are thus complicated by the effect of the ions of each component in the solution upon the value of $(\mathrm{cmc})_i^{\circ}$ of the other component: correction for this has to be made for each mixture before equation 10-1 is applied.

Amphiphiles with different head groups form mixed micelles as readily as amphiphiles with the same head group (Shinoda et al., 1963; Becher, 1967). Ideal mixing in the micelle is not to be expected in this case, but ideal dilute solution behavior may be anticipated when one component is in large

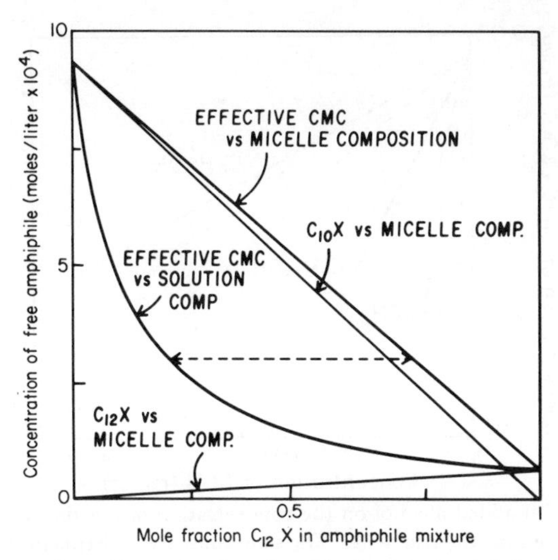

Fig. 10-1. Thermodynamic properties of an ideal mixture of dodecyl hexaoxyethylene glycol monoether ($C_{12}X$) with the corresponding decyl compound ($C_{10}X$). The figure gives the concentrations of free $C_{12}X$ and $C_{10}X$ as a function of micelle composition and the sum of these quantities (designated "effective cmc"). The effective cmc is also shown as a function of the composition of free amphiphile in solution. Horizontal lines between the two latter curves provide the composition of free amphiphile in solution in equilibrium with any given composition in the micelle, as is illustrated by the dashed line, which shows that a mixture of free amphiphile in solution containing 80% $C_{10}X$ and 20% $C_{12}X$ is in equilibrium with a micelle containing 28% $C_{10}X$ and 72% $C_{12}X$.

excess, that is, equation 10-1 would apply to the component in excess and the analog of Henry's law would apply to the other component,

$$X_{j,W} = KX_{j,\text{mic}} \tag{10-2}$$

where K may be smaller or larger than $(\text{cmc})_j^\circ$ depending on the magnitude of the repulsive interaction between the different head groups as compared to that between identical head groups in the pure micelles.

Equation 10-2 would also apply to the incorporation of small amounts of hydrophobic or amphiphilic substances that themselves do not form micelles in micelles of another amphiphile. Combining equation 10-2 with equation 10-1 for the principal component of the micelle leads to the prediction that the addition of any such substance to the solvent medium should lead to a decrease in $X_{i,W}$ and that this decrease should be linearly related to $X_{j,W}$ when the latter is small. An example of this effect is provided by Fig. 10-2.

It is of interest that the slopes of the lines in Fig. 10-2 are considerably

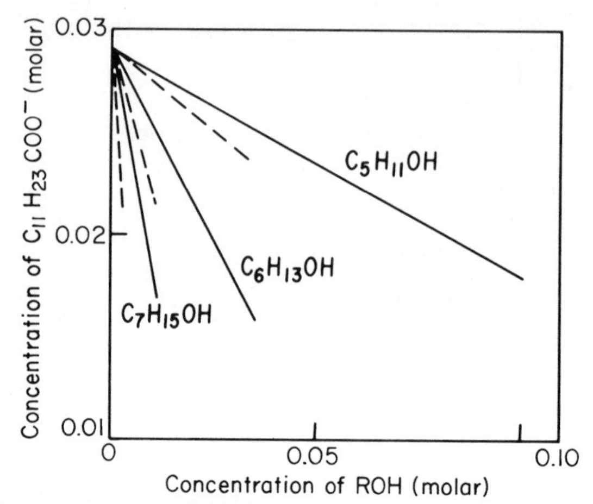

Fig. 10-2. Effect of added alcohol on the concentration of free dodecanoate in equilibrium with its micelles in water at 10°C. The solid lines are experimental lines taken from Shinoda et al., (1963). The dashed lines are calculated with the assumption that μ°_{mic} of an alcohol in the micelle is equal to μ°_{ROH} of pure liquid alcohol.

smaller than the slopes one would calculate on the basis of the assumption that the alcohol molecules in the micelle are at the same chemical potential as they would be in an ideal liquid solution of alcohols. The slopes expected on the basis of this assumption can be calculated by use of equation 3-3, and are shown as dashed lines in Fig. 10-2. This result is in accordance with the experimental observation that the aliphatic alcohols do not form micelles in aqueous solution, so that the cmc for micelle formation must be larger than the concentration in solution at which phase separation occurs. Nonideality in the mixed micelle may also have an effect, of course but, if this is ignored, the data indicate that $\mu^\circ_{\text{mic}} - \mu^\circ_W$ for formation of hypothetical ROH micelles is about 400 cal/mole more positive than $\mu^\circ_{\text{ROH}} - \mu^\circ_W$ as given by equation 3-3.

It should be noted that virtually all substances containing hydrophobic moieties decrease the concentration of an amphiphile in equilibrium with its micelles. For small molecules (e.g., ethanol) this is generally expressed by stating that their addition to the solvent decreases the cmc. This decrease can be explained only by the incorporation of the added substance into the micelle, since simple addition of the alcohol and similar substances to the solvent must diminish the hydrophobic effect, that is, it would decrease μ°_W for the micelle-forming amphiphile and (with no effect on μ°_{mic}) would therefore *increase* the cmc. In fact an increase in cmc values is observed upon addition of polar molecules, such as urea, which are known to diminish the hydrophobic effect but which could not be incorporated in a micelle with a hydrophobic core (Emerson and Holtzer, 1967).

The formation of mixed micelles and the incorporation of small molecules with hydrophobic groups into micelles, and the fact that experimental observations on these phenomena are in reasonable agreement with predictions based on treatment of the micelle as a liquid mixture, attest to the liquidlike nature of the micelle interior. These phenomena also provide the basis for the detergent action of amphiphile micelles.

REFERENCES

Becher, P. (1967). In *Nonionic Surfactants*, M. J. Schick, Ed., Marcel Dekker, Inc., New York.

Emerson, M. F., and A. Holtzer. (1967). *J. Phys. Chem.*, **71**, 3320.

Hall, D. G., and B. A. Pethica. (1967). In *Nonionic Surfactants*, M. J. Schick, Ed., Marcel Dekker, Inc., New York.

Shinoda, K., T. Nakagawa, B. Tamamushi, and T. Isemura. (1963). *Colloidal Surfactants*, Academic Press, New York.

ELEVEN

MONOLAYERS

Amphiphilic molecules or ions in an aqueous solution respond to the hydrophobic effect in another way, by migration to the surfaces of the aqueous medium to form oriented monolayers in which the polar head groups remain in the aqueous medium while the hydrophobic moieties are expelled from it. Such monolayers will generally be present at the water–air interface of any aqueous solution of amphiphiles, and are readily detected by a marked reduction in the surface tension. They will also form at the surface of the container in which the solution is placed, or at the interface between an aqueous solution and an immiscible organic liquid with which it may be placed in contact.

The study of surface monolayers holds an honorable place in the history of science. Surface films can be formed not only by migration from the bulk solution to the surface, but also by the spontaneous spreading of "insoluble" or sparingly soluble liquids placed on a water surface. When this is done with a pure hydrocarbon oil, no spreading occurs, for in the absence of a hydrophilic group there is no driving force for intimate association between the deposited substance and the underlying water (Hardy, 1912). When the oil is amphiphilic, however, rapid spreading occurs as each surface molecule seeks to place its hydrophilic part in the water. The result is the formation of a film that is only one molecule thick. It is of interest to note that the remarkable thinness of such films was observed as early as 1765 by Benjamin Franklin, who spread a known amount of olive oil on a pond at Clapham Common in London and, from the area covered, estimated that the film had to have a thickness of no more than 10^{-7} inches, or 25 Å (Davies and Rideal, 1961, p. 217). A similar conclusion was reached later by Rayleigh (1890). Subsequent studies of considerable historical importance are those of Traube (1891), which were summarized in Chapter 1, and the work of Langmuir (1917) on the orientation of molecules or ions in surface layers. It is this unique orientation that is responsible for the dramatic decrease in the surface tension of water by surface-active agents, since the surface of a saturated film has the weak intermolecular cohesion characteristic of hydrocarbons instead of the strong cohesion between water molecules.

There are several excellent textbooks on surface chemistry (Adam, 1941; Adamson, 1967; Davies and Rideal, 1961; Harkins, 1952), which provide a wealth of theoretical and experimental information on the subject. Only brief mention of aspects that are of interest in connection with the formation of micelles or membranes will be made here and in later chapters.

FREE ENERGY OF TRANSFER TO AN AIR-WATER INTERFACE

The standard unitary free energy of transfer of amphiphiles from the bulk of a solution to the surface can be calculated from the data of Traube (1891), using the same relation as has been used previously for other transfer processes. Ideality may be assumed both in the solution and in the surface film, since the results extend to low concentrations where only a fraction of the available surface is occupied, and can be extrapolated to infinite dilution (Langmuir, 1917). Under these conditions

$$\overset{\circ}{\mu}_{\text{surf}} - \overset{\circ}{\mu}_{W} = RT \ln X_W / X_{\text{surf}} \qquad (11\text{-}1)$$

To evaluate the concentration in the surface film (X_{surf}) one has to know how many molecules of amphiphile are in the surface layer and also how thick this layer is, that is, how many water molecules it contains. It is reasonable to assume, in a homologous series, that lowering of surface tension is a measure of X_{surf}, so that *relative* values of $\overset{\circ}{\mu}_{\text{surf}} - \overset{\circ}{\mu}_{W}$ can be obtained simply from the values of X_W that produce equal lowering of the surface tension. Applying this principle to Traube's results, Langmuir (1917) showed that all of the results, for homologous series of fatty acids, esters, alcohols, and so on, could be represented (in cal/mole) as

$$\overset{\circ}{\mu}_{\text{surf}} - \overset{\circ}{\mu}_{W} = \text{constant} - 625 n_{\text{C}} \qquad (11\text{-}2)$$

with only the constant term dependent on the head group.

The result given by equation 11-2 is qualitatively consistent with the data for transfer from water to a hydrocarbon solvent given in Chapters 2 and 3. The increment in $\overset{\circ}{\mu}_{HC} - \overset{\circ}{\mu}_{W}$ per carbon atom is about $-$ 850 cal/mole, and the difference between this figure and the increment of -625 cal/mole in equation 11-2 is to be ascribed to the weak cohesive forces between hydrocarbon chains in a liquid medium. When similar calculations are made for adsorption of amphiphiles to a water–paraffin oil interface (Davies and Rideal, 1961; p. 158) one obtains an increment in $\mu^{\circ}_{surf} - \mu^{\circ}_{W}$ of -810 cal/mole per CH_2 group, which is much closer to the value for transfer to a bulk hydrocarbon medium. The residual difference that remains may be ascribed to loss of entropy resulting from orientation of the amphiphile molecules at the interface.

PRESSURE-AREA CURVES

Monolayers at an interface can exist in different physical states, analogous to the gaseous, liquid, and solid states of matter in bulk. Information regarding these states and the transitions between them can be obtained by measuring the *surface pressure* of a monolayer as a function of the surface concentration. The surface pressure (Π) is defined as the lateral pressure that must be applied to prevent the film from spreading. The surface concentration may be measured directly on the basis of how much amphiphile has been deposited on the surface. For substances that have a readily measurable solubility in the aqueous solution, the surface concentration may also be determined by means of the Gibbs adsorption equation, which rigorously relates the surface concentration to the effect of bulk concentration on the surface tension. Measurements are made on a large planar surface, and are often reported in terms of plots of Π versus the available area per molecule (A), which is the reciprocal of the surface concentration. This type of plot is particularly useful at high surface pressures, where it defines the minimum area to which a molecule at the surface can be compressed. Illustrative plots are shown in Figs. 11-1 and 11-2.

When the surface is sparsely covered, surface monolayers are "gaseous," that is, they consist of individual adsorbed molecules moving independently. At great dilution their behavior is analogous to an ideal gas and the two-dimensional analog of the ideal equation of state

$$\Pi A = kT \tag{11-3}$$

is obeyed, with k representing Boltzmann's constant. As the pressure is increased, gaslike films become nonideal, and an equation of state analogous to the van der Waals equation for bulk gases applies, with corrections for the surface area from which an individual adsorbed molecule excludes all others and for the existence of weak attractive forces between molecules.

At an air–water interface the gaslike state is not stable except at very low pressures. As the pressure is increased, the cohesive force between the hydrocarbon chains leads to condensation of the gaslike film to a liquidlike film, as

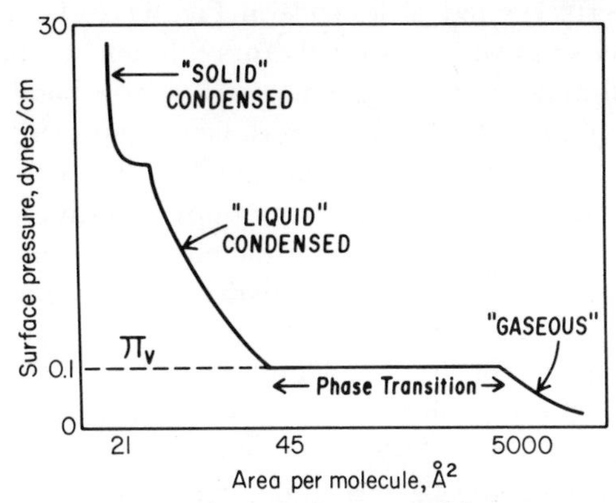

Fig. 11-1. Schematic representation of phase transitions at an air-water interface. Adapted from Harkins (1952). Pressure and area scales are not linear and numerical values are intended only to indicate typical orders of magnitude.

illustrated by Fig. 11-1. This liquidlike film, unlike the bulk liquid state, is highly compressible. The reason for this is that the hydrocarbon chains of the amphiphile can form a coherent film while lying flat on the surface and, with increase in the applied surface pressure, can change their orientation until they are perpendicular to the surface, at a molecular area many times smaller than that of the most expanded liquidlike film. The compression process is not necessarily continuous and two or more distinct "liquid" phases and transitions between them can often be observed (Harkins, 1952, p. 106). At high compression a further transition to a solidlike state may be observed, in which the adsorbed molecules become immobile: this generally requires the existence of an attractive force between head groups in addition to the force of cohesion between hydrocarbon chains.

The situation at an interface between an aqueous solution and an immiscible organic liquid, such as a liquid hydrocarbon, is much simpler. Here the cohesive force between hydrocarbon chains is satisfied at all surface concentrations and the surface monolayer remains gaslike to high pressures unless there are attractive forces between head groups, or unless one chooses for the organic phase a substance with which the hydrocarbon chains are not completely miscible.

Most experimental studies have been made at water–air surfaces, and have involved relatively high surface concentrations. Many such studies have employed amphiphile molecules with small head groups that do not repel one another, and therefore do not form micelles in the bulk solution, such as undissociated fatty acids. The results primarily reflect the association tendencies and orientations of hydrocarbon chains, as illustrated for example by the data for stearic and oleic acids in Fig. 11-2. They show that long saturated alkyl chains condense directly (at room temperature) to a solidlike incompressible film, whereas the presence of a double bond, as in oleic acid, leads to a relatively expanded compressible liquidlike film. The limiting surface area at high pressures, for all amphiphiles with single saturated alkyl chains and sufficiently small head groups is always found to be about 21 $Å^2$ per molecule, that is, the same as the minimum area per hydrocarbon chain calculated for a bilayer when the hydrocarbon chains are perpendicular to the bilayer surface (Fig. 9-1). Liquid films formed by amphiphiles containing mutually repelling head groups are more expanded than the monolayers formed by amphiphiles in which such repulsion is minimal or absent, as illustrated in Fig. 11-2 for an alkyl trimethylammonium salt.

In connection with micelle formation, the greatest interest attaches to data obtained at a hydrocarbon–water interface, because these data provide a direct measure of the repulsive force that is the size-limiting factor in micelle formation. Illustrative data for an anionic and a cationic amphiphile are given in Fig. 11-2. Brooks and Pethica (1965) have shown that these curves are independent of the alkyl chain length, as is to be expected. (This is of course not true for similar data at an air-water interface because the pressure-area curves in that case depend on the van der Waals attraction between hydrocarbon chains as well as on the repulsion between head groups.) An interesting feature of the results of Brooks and Pethica is the

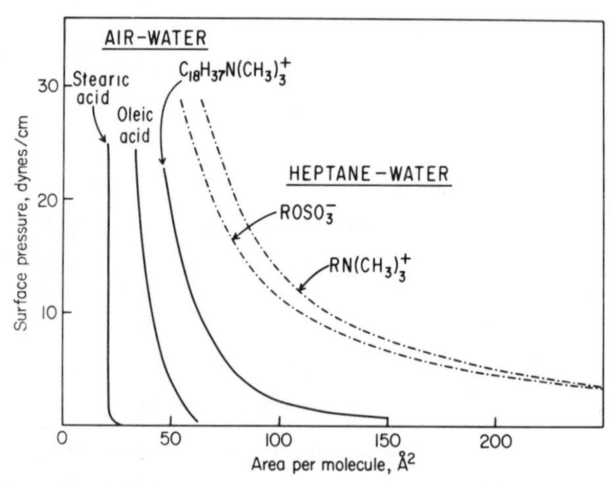

Fig. 11-2. Pressure-area curves at an air-water interface (Davies and Rideal, 1961) and at a heptane-water interface (Brooks and Pethica, 1965). The aqueous phase contained 0.1 to 0.5 *M* NaCl in the experiments with ionic amphiphiles. The results at the heptane-water interface are essentially independent of alkyl chain length from $n_C = 12$ to $n_C = 22$.

difference between the curves for $RN(CH_3)_3^+$ and $ROSO_3^-$, which shows that the repulsive force between the cationic head groups is somewhat greater than the repulsive force between sulfate head groups. This result is consistent with the observation that under otherwise identical conditions $RN(CH_3)_3^+$ forms smaller globular micelles than $ROSO_3^-$ (e.g., Emerson and Holtzer, 1967), which, as shown by Fig. 9-1, means that the head groups tend to be farther apart.

Experiments such as these can potentially provide a great deal of information relevant to micelle formation. It should for example be possible to estimate the repulsive component W°_{mic} of the free energy of micelle formation (equation 7-12) from the results at a hydrocarbon-water interface. This and other possible correlations have not however been attempted so far. One of the reasons for this is that the determination of pressure-area curves requires extraordinary care, because trace impurities are often concentrated at an interface, so that levels of purity in the bulk of a solution that are satisfactory for all other purposes do not guarantee the absence of impurities at an interface.

The Transition from Gaslike to Liquidlike States

Gershfeld and Pagano (1972) have recently developed a film balance that makes it possible for the first time to measure very low surface pressures and thereby to investigate the transition from gaslike to expanded liquidlike or condensed surface films. Initial measurements reported by Gershfeld and Pagano include values for the surface pressure at which the transition occurs, which is analogous to the equilibrium vapor pressure of a bulk liquid or solid, and which they call the "surface vapor pressure," Π_v. By measuring Π_v at different temperatures, values for ΔH and ΔS for the surface phase transition were determined. It was observed that ΔS for formation of a condensed film (e.g., stearic acid in Fig. 11-2) is considerably more negative than ΔS for formation of an expanded film, presumably reflecting the ordering process in formation of a solidlike structure. The values of ΔS for formation of an expanded film are more positive than one would have expected if contributions from changes in water structure are not taken into consideration. Presumably this reflects the disordering that accompanies displacement of water molecules from the surface, as discussed in Chapter 5.

Other Physical Measurements

In addition to surface pressure, other physical properties of surface films have been studied as a function of surface concentration. They include measurements of the electrical potential across a monolayer, which provides information on the orientation of the adsorbed molecules; measurements of surface viscosity; and optical measurements. The techniques employed and the available results are described in the references given at the beginning of this chapter.

MIXED MONOLAYERS

Monolayers, like micelles, may contain mixtures of amphiphiles. Pressure-area curves of such mixed films have to be interpreted with great caution, because more than one liquidlike phase of different composition may be present at the same time (Gaines, 1966). An additional difficulty, which applies to all surface films, but is an especially important possible source of error for mixed monolayers, is that equilibration between the monolayer and the bulk solution may be very slow. For mixed monolayers the equilibration process may influence the monolayer composition, which may therefore change with time in an experiment in which (as is often done) the monolayer constituents are spread upon the surface without stirring or other means of accelerating the equilibration process.

Experimental pressure-area data for mixed liquidlike films have been obtained at air–water surfaces, and they indicate that interactions between surface components are often far from ideal. Both "condensation" and "expansion" have been observed, that is, the area per adsorbed molecule may be less than or greater than the mean area calculated for the same pressure on the basis of the pressure-area curves of the pure components. Since orientation of the hydrocarbon chains is the primary determinant of surface area (except at the high pressure limit) it is not surprising to find that the lengths of alkyl chains have an important influence on the observed results.

It has been suggested (Pagano and Gershfeld, 1972) that measurements of the equilibrium surface vapor pressure (Π_v) at the transition between gaseous and liquidlike films provides a more readily interpretable means of investigating mixed monolayers. This phenomenon obeys the same thermodynamic laws as apply to the vapor pressures of bulk solutions. Pagano and Gershfeld have shown that some mixtures behave ideally by this criterion and others show positive deviations from ideality. Negative deviations, corresponding to a greater mutual attraction between disparate molecules than between the molecules of the pure components, were not observed in the limited number of systems they investigated. Some of the systems studied by Pagano and Gershfeld gave Π_v versus composition curves indicative of incompletely miscibility of the components, that is, two liquidlike phases appeared to coexist at equilibrium over some part of the composition range. The latter result reinforces the cautionary note expressed above with regard to pressure-area curves: their interpretation is not clear if two distinct liquidlike phases are present.

INTERACTIONS OF SOLUBLE SUBSTANCES WITH MONOLAYERS

Monolayers provide a convenient system for the study of the interaction between water-soluble substances and amphiphiles. A trivial example is provided by the effects of dissolved salts on the pressure-area curves of amphiphiles with ionic head groups. The system is particularly useful for the study of the interaction between dissolved proteins and monolayers formed by biological lipids and some examples of its use will be cited later in this book.

REFERENCES

Adam, N. K. (1941). *The Physics and Chemistry of Surfaces*, 3rd. ed., Oxford University Press.

Adamson, A. W. (1967). *Physical Chemistry of Surfaces*, 2nd. ed., Wiley-Interscience, New York.

Brooks, J. H., and B. A. Pethica. (1965). *Trans. Faraday Soc.*, **61,** 571.

Davies, J. T., and E. K. Rideal. (1961). *Interfacial Phenomena*, Academic Press, New York.

Emerson, M. F., and A. Holtzer. (1967). *J. Phys. Chem.*, **71,** 1898.

Gaines, G. L., Jr. (1966). *Insoluble Monolayers at Liquid-Gas Interfaces*, Wiley-Interscience, New York.

Gershfeld, N. L., and R. E. Pagano, (1972). *J. Phys. Chem.*, **76,** 1231.

Hardy, W. B. (1912). *Proc. Roy. Soc.*, **A86,** 610.

Harkins, W. D. (1952). *The Physical Chemistry of Surface Films*, Reinhold, New York.

Langmuir, I. (1917). *J. Am. Chem., Soc.*, **39,** 1848.

Pagano, R. E., and N. L. Gershfeld. (1972). *J. Phys. Chem.*, **76,** 1238.

Rayleigh, Lord (1890). *Proc. Roy. Soc.*, **47,** 364.

Traube, J. (1891). *Ann.*, **265,** 27.

TWELVE

BIOLOGICAL LIPIDS

The biochemist uses the term "lipid" to define all organic molecules of biological origin that are highly soluble in organic solvents and only sparingly soluble in water. Included in this definition are fats, some hormones and vitamins, and many other substances of diverse chemical identity and biological function. The discussion in this chapter will be confined to lipids present in biological membranes or in soluble lipoprotein complexes, and they fall into two classes:

1. Amphiphile molecules or ions of the type discussed previously in this book, containing a strongly hydrophilic head group and one or more (most often *two*) long hydrocarbon tails.

2. Cholesterol and acyl esters of cholesterol. These substances do not resemble any we have previously considered: their principal constituent is pure hydrocarbon in the form of the relatively rigid steroid ring. Cholesterol, the structural formula for which is

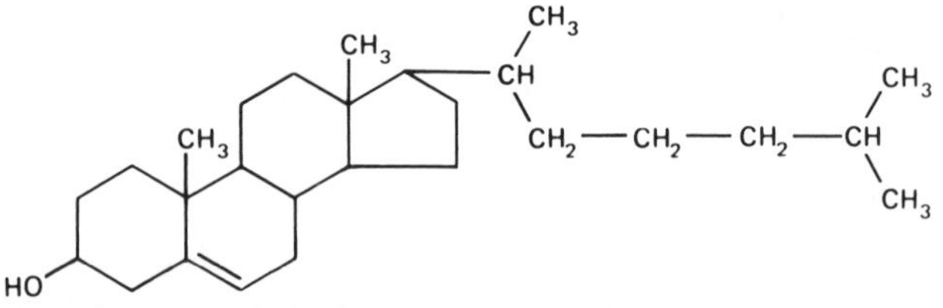

may be considered amphiphilic by virtue of possession of a single OH group. In cholesteryl esters this OH group forms an ester link with a fatty acid. In either case the affinity for water is very weak compared to that of the strongly hydrophilic head groups of other membrane lipids.

The most abundant lipids in the first category contain two hydrocarbon chains attached to a single polar head group. Among lipids of this type are most of the phosphoglycerides, which have the general formula

$$\begin{array}{lccccll} & & & & & \mathrm{O^-} & \\ & & & & & | & \\ & \mathrm{CH_2} & - & \mathrm{CH} & -\mathrm{CH_2}-\mathrm{O}- & \mathrm{P} & -\mathrm{O}-\mathrm{X} \\ & | & & | & & \| & \\ & \mathrm{O} & & \mathrm{O} & & \mathrm{O} & \\ & | & & | & & & \\ \mathrm{O}= & \mathrm{C} & & \mathrm{C}=\mathrm{O} & & & \\ & | & & | & & & \\ & \mathrm{R_1} & & \mathrm{R_2} & & & \end{array} \qquad \text{(I)}$$

where R_1 and R_2 are hydrocarbon chains. The head group bears a net negative charge at neutral pH if the X group in this formula is neutral or zwitterionic, as follows:

Phosphatidic acid	$-X = -H$
Phosphatidylserine	$-X = -CH_2CH(NH_3^+)COO^-$
Phosphatidylinositol	$-X = -C_6H_6(OH)_5$
Phosphatidylglycerol	$-X = -CH_2CHOHCH_2OH$

Alternatively the head group may be zwitterionic at neutral pH, if the X group bears a positive charge, as in

Phosphatidylethanolamine	$-X = -CH_2CH_2NH_3^+$
Phosphatidylcholine	$-X = -CH_2CH_2N(CH_3)_3^+$

Aminoacyl derivatives of phosphatidyl glycerol occur in some bacteria (Houtsmuller and Van Deenen, 1965; Lennarz, 1972). Included in this category is the lysyl ester of phosphatidylglycerol, which bears two positive charges, conferring a net positive charge on the head group as a whole. Apart from this instance, phosphoglycerides with positively charged head groups are rare or nonexistent.

Plasmalogens represent a variant form of I in which one of the ester links is replaced by an ether linkage, and similar molecules with two ether linkages have also been observed. Another lipid, chemically quite distinct but geometrically similar, is sphingomyelin

$$\begin{array}{lccccll} & & & & & \mathrm{O^-} & \\ & & & & & | & \\ \mathrm{HO}- & \mathrm{CH} & - & \mathrm{CH} & -\mathrm{CH_2}-\mathrm{O}- & \mathrm{P} & -\mathrm{O}-\mathrm{CH_2CH_2N(CH_3)_3^+} \\ & | & & | & & \| & \\ & \mathrm{CH} & & \mathrm{NH} & & \mathrm{O} & \\ & \| & & | & & & \\ & \mathrm{HC} & & \mathrm{C}=\mathrm{O} & & & \\ & | & & | & & & \\ & \mathrm{R_1} & & \mathrm{R_2} & & & \end{array} \qquad \text{(II)}$$

Most glycolipids are also sphingolipids, with the general formula

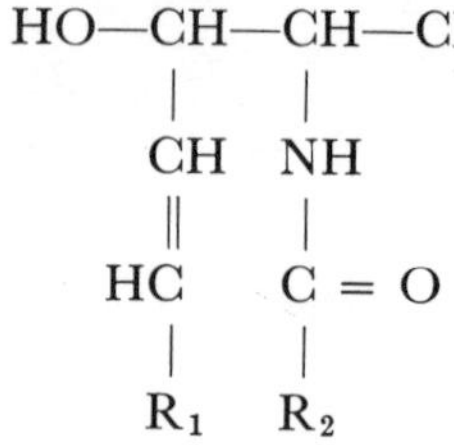

$$\begin{array}{lclcl}
\mathrm{HO{-}} & \mathrm{CH} & \mathrm{{-}} & \mathrm{CH} & \mathrm{{-}CH_2{-}O{-}X} \\
 & \vert & & \vert & \\
 & \mathrm{CH} & & \mathrm{NH} & \\
 & \Vert & & \vert & \\
 & \mathrm{HC} & & \mathrm{C=O} & \\
 & \vert & & \vert & \\
 & \mathrm{R_1} & & \mathrm{R_2} &
\end{array} \qquad \text{(III)}$$

The carbohydrate moiety X may be a single neutral hexose sugar (as in cerebrosides), a sulfated sugar (cerebroside sulfate or sulfatide), or a more complex oligosaccharide. The head group of sulfatides bears a negative charge, as does the head group of gangliosides, in which X is a complex oligosaccharide containing one or more moles of sialic acid.

Mitochondria contain considerable quantities of cardiolipin, in which *four* hydrocarbon chains are attached to a single head group. This lipid is essentially a dimeric form of I. Two molecules of the diacyl glycerophosphate moiety are linked by one molecule of glycerol (i.e., —X is replaced by $—CH_2CHOHCH_2^-$). There are two negative charges on the head group.

Lipids with a single hydrocarbon chain per head group occur in only very small quantities.[1] Lipids in this category are the lysophospholipids (formula I with one of the acyl group removed) and the free fatty acid anions. As noted in Chapter 10, all hydrophobic or amphiphilic molecules tend to be incorporated in micelles formed by other amphiphiles and, considering membranes tentatively as having some resemblance to micelles formed by the principal membrane lipids, the presence of some minor components, such as fatty acid anions, may simply be a reflection of their presence in the medium with which the membrane is in contact.

Analytical data for the lipid content of several membranes are shown in Table 12-1. They are intended primarily to show that composition with respect to lipid classes is quite different for different membranes. The common features of all the data are (1) that 75% or more of membrane lipid consists of amphiphilic molecules with two hydrocarbon chains per head group, (2) that about 20 to 30% of the lipid contains anionic head groups, and (3) that there are no cationic head groups at all, those that are not anionic being neutral or zwitterionic. (As noted above, positively charged head groups do occur in some bacterial membranes.)

[1] The membrane of the hormone-secreting granules of the adrenal cortex may constitute an exception; it has been shown to contain about 15% of lysophosphatidylcholine (Winkler and Smith, 1968). A similar high content of lysophosphatidylcholine in zymogen-secreting granules of the pancreas has however been shown to be an artifact, arising from enzymatic degradation during the isolation procedure (Meldolesi et al., 1970).

Table 12-1. Lipid Compositions of Some Biological Membranes[a] (expressed as percent by weight of total lipid)

	Human Erythrocyte[d]	Human Myelin[e]	Beef Heart Mitochondria[f]	E. Coli[g]
Phosphatidic acid	1.5	0.5	0	0
Phosphatidylcholine[b]	19	10	39	0
Phosphatidylethanolamine[b]	18	20	27	65
Phosphatidylglycerol	0	0	0	18
Phosphatidylinositol	1	1	7	0
Phosphatidylserine	8.5	8.5	0.5	0
Cardiolipin	0	0	22.5	12
Sphingomyelin	17.5	8.5	0	0
Glycolipids	10[h]	26[i]	0	0
Cholesterol[c]	25	26	3	0

[a] Quantitative lipid analysis is a rapidly developing science, and many of these figures are likely to change as it progresses. In bacterial membranes, lipid composition depends to some extent on growth phase and nutritional factors.
[b] Up to one-third of the phosphatidyl derivatives may be in the plasmalogen form, that is, with an ether rather than an ester link between the hydrocarbon chain and the glyceryl moiety. For our purposes the distinction is not important.
[c] The figures are for free cholesterol. Cholesterol esters occur in soluble lipoproteins, but have not been found in membranes.
[d] Rouser et al. (1968).
[e] Dickerson (1968), Rouser et al. (1968). The values given are for central nervous system myelin.
[f] Rouser et al. (1968). The values given are for total mitochondrial lipid. Inner and outer membranes differ significantly in composition.
[g] Ames (1968).
[h] 80% of the glycolipid is cerebroside, 20% is sulfatide.
[i] Principally polyhexosides and gangliosides.

The hydrocarbon chains of biological lipids are invariably very long. Only small amounts of fatty acids with fewer than 16 carbon atoms are usually obtained upon hydrolysis of diacyl lipids, and chain lengths to 24 carbon atoms occur frequently. An almost universal feature is the presence of unsaturated fatty acids,[2] typically to the extent of about 50% of the total

[2] In some bacterial lipids a methylene group may be added across the unsaturated bond to produce a cyclopropyl group, or branched saturated fatty acids may under certain conditions replace unsaturated fatty acids. The important factor is geometrical and not chemical: about half the acyl chains must have structures that prevent them from forming parallel close-packed arrays with linear saturated hydrocarbon chains.

fatty acid content, although there are exceptions. For example, membranes of lung tissues contain hydrocarbon chains derived predominantly from saturated fatty acids, dipalmitoyl phosphatidylcholine being the principal constituent (Abrams, 1966). The fatty acid composition of the microorganism *Mycoplasma laidlawii B* can be altered dramatically by regulation of the growth medium, and the organism is viable with a saturated fatty acid content of as much as 90% (McElhaney and Tourtelotte, 1969). In the normal situation, where the number of saturated and unsaturated hydrocarbon chains is about equal, one chain of each kind is often present on the same molecule: in the phosphoglycerides (formula I) of many species, R_1 tends to be a saturated fatty acid and R_2 an unsaturated fatty acid. Unsaturated fatty acids may contain up to six double bonds. Double bonds in unsaturated fatty acids from bacterial or animal membranes are usually in the *cis*-configuration and, in polyunsaturated acids, tend not to be conjugated. Reviews by Rouser et al. (1968) and Hill and Lands (1970) may be consulted for additional information.

All of the biological lipids described above, except cholesterol and its esters, resemble the simpler amphiphiles discussed in preceding chapters. They are expected to form micelles in aqueous media and, because their constituent hydrocarbon chains are invariably very long, the concentration of unassociated lipid in equilibrium with the micelles (which we shall continue to assume is essentially the same as the critical micelle concentration) should be very small. For the predominant type of molecule, containing two hydrocarbon chains per head group, the micelle is expected to be of the bilayer type, and there should be a tendency for the bilayers to form closed vesicles (Chapter 9). Mixtures of lipids should readily form mixed micelles and, if most of the constituent molecules contain two hydrocarbon chains per head group, the mixed micelles should also be bilayers. All available experimental data, some of which will be summarized in this chapter, support these predictions.

In considering the physicochemical data to be presented we should keep in mind the likelihood, as discussed in Chapter 6, that micellar structures formed by amphiphile molecules with very long *saturated* hydrocarbon chains may have hydrophobic cores in which the hydrocarbon chains are in an ordered paracrystalline array at room temperature, instead of being liquid. For molecules with *unsaturated* hydrocarbon chains, on the other hand, the melting point for ordered structures is expected to be below room temperature, and the hydrophobic cores may accordingly be in a fluid state. The data to be presented in this and the following chapter will in fact demonstrate that ordered and disordered hydrophobic cores exist much as expected, and the results of the following chapter will permit us to refine our concepts of the meaning of "order" and "disorder" in these systems.

Many of the results to be discussed were obtained with lipids isolated from natural sources, containing molecules with a mixture of hydrocarbon chains. The material of this type that has been employed more frequently than any other is egg yolk phosphatidylcholine. This material is pure with regard to the head group, but contains a mixture of acyl chains, about 50% of them saturated and 50% unsaturated. Principal constituents after hydrolysis of the ester linkage were found in a typical analysis (Huang et al., 1964) to be palmitic acid, stearic acid, oleic acid (one double bond) and linoleic acid (two double bonds). A small amount of $C_{19}H_{32}COOH$ (four double bonds) was also present. A typical individual molecule in the mixture probably contains one saturated and one unsaturated hydrocarbon chain, as mentioned earlier. Ordered arrays of the hydrocarbon chains have low stability in a mixture of this kind, and egg phosphatidylcholine will be seen to resemble pure synthetic molecules with unsaturated hydrocarbon chains in that it forms bilayers with a liquid hydrophobic core at room temperature. In terms of a number of physical properties not discussed here, the egg yolk product resembles pure dioleyl phosphatidylcholine (DeGier et al., 1968).

CRITICAL MICELLE CONCENTRATION

Because the cmc values are so small, they are difficult to measure. Experimental values are available for palmitoyl lysophosphatidylcholine, which contains only one hydrocarbon chain, and for dipalmitoyl phosphatidylcholine. They are shown in Table 12-2. This table also shows predicted values, based on extrapolation of the data shown in Figs. 7-2 and 7-3. Since no data are available for simple amphiphiles containing the phosphatidylcholine head group, we have used the results for the N-alkyl betaines which, like the phosphatidylcholines, contain a zwitterionic head group, as a suitable model. Predicted values for the cmc would have been lower by a factor of 10 if the alkyl glucosides had been used as a basis for the extrapolation. For the dipalmitoyl derivative we have assumed that the second hydrocarbon chain exerts an effect equivalent to about 60% of that which would result from extending a single chain by the same number of carbon atoms, as Fig. 7-3 suggests. This leads to the prediction that the cmc of dipalmitoyl phosphatidylcholine should be about the same as the cmc of the corresponding lysophosphatidylcholine with a C_{24} saturated hydrocarbon chain. The predictions made in this way are seen to be in excellent agreement with the experimental results, which suggests that the cmc of most pure lipids of this type can probably be estimated reasonably well on the basis of the results for simpler amphiphiles.

Table 12-2. Critical Micelle Concentrations of Biological Lipids in Aqueous Solution at 25°C

	cmc (moles/liter)	cmc (mole fraction)	$\mu^\circ_{mic} - \mu^\circ_W$ (kcal/mole)[a]
Palmitoyl[b] *Lysophosphatidylcholine*			
Observed[c]	$<12 \times 10^{-5}$	$<2.1 \times 10^{-6}$	<-7.7
Calculated[d]	6×10^{-5}	1.0×10^{-6}	—
Dipalmitoyl[b] *Phosphatidylcholine*			
Observed[e]	4.7×10^{-10}	8.4×10^{-12}	-15.1
Calculated[d]	11×10^{-10}	21×10^{-12}	—

[a] Using equation 7-3 with $f_w = 1$.
[b] The palmitoyl moiety contains a C_{15} hydrocarbon chain.
[c] Lewis and Gottlieb (1971). Robinson and Saunders (1958) observed a cmc in the range of 2 to 20 $\times 10^{-5}$ *M* for a lysophosphatidylcholine with mixed hydrocarbon chains.
[d] Based on extrapolation of the data of Figs. 7-2 and 7-3, by using the betaine head group as representative of the phosphatidylcholine head group. See text for details.
[e] Smith and Tanford (1972).

The possibility that the hydrocarbon chains of dipalmitoyl phosphatidylcholine may actually be in an ordered paracrystalline array instead of in a liquid core should not have a large effect on the cmc. The free energy of melting of the ordered array is probably less than 1 kcal/mole and the difference between the cmc of otherwise identical micelles with ordered and liquid cores should thus be less than a factor of 5. This difference is not much greater than the precision with which the cmc can be measured: the probable error is about a factor of two.

SOLUBLE MICELLES

The micelles formed by lysophosphatidylcholines are small and globular, similar to those observed for simpler single-chain amphiphiles. The value for $\bar{m}$ is about 180, both for synthetic palmitoyl lysophosphatidylcholine (Lewis and Gottlieb, 1971) and for a mixed product obtained by enzymatic cleavage of egg yolk phosphatidylcholine (Saunders, 1966). The intrinsic viscosity of the latter was found to be 4 cc/g. Because of the virtual absence of phospholipids with a single hydrocarbon chain in biological membranes, micelles of this type are not of much interest, and little work has been done on them.

Phospholipid molecules containing two hydrocarbon chains are relatively difficult to disperse in aqueous solution in micellar form. When water is added to dry phospholipid, swelling takes place, with formation of liquid crystalline phases of different kinds (see below). Further addition of water tends to lead to dispersal of this lipid without disruption of the ordered aggregated structures. This is probably due to the tendency of bilayers to form closed structures in the presence of water. Electron micrographs of phospholipid preparations containing water or of discrete "liposomes" obtained from them (Dervichian, 1964; Bangham et al., 1965) invariably show multilayered closed structures, such as illustrated by Fig. 12-1, which cannot be further subdivided without rupture of the individual layers.

As was first observed by Saunders et al. (1962) small soluble micelles can be obtained from multilayered structures by ultrasonic irradiation. The micelles so formed represent a mixture of vesicles, some still bounded by several bilayers. A fraction containing only vesicles bounded by a single bilayer can however be separated from such a mixture by gel exclusion chromatography. A detailed study of such vesicles, derived from egg yolk phosphatidylcholine, has been reported by Huang (1969). They were found to be quite uniform in size, with a molecular weight of 2.1×10^6, corresponding to $\bar{m} = 2680$. Electron micrographs, the Stokes radius as measured by gel chromatography and an intrinsic viscosity of 4.1 cc/g are all consistent with a roughly spherical vesicle, with an external diameter of about 250 Å and an internal cavity with a diameter of about 140 Å containing about 0.6 g solvent (mostly water) per g of phospholipid. Both the thickness of the bilayer and the area occupied per head group in these vesicles are consistent with results obtained from X-ray diffraction of liquid crystalline egg phosphatidylcholine (see below).

Similar vesicles, bounded by a single bilayer, but less uniform and of somewhat larger average size, have been obtained from soy bean lipids (Miyamoto and Stoeckenius, 1971), as illustrated by Fig. 12-1*b*. Clear dispersions can be obtained by ultrasonic irradiation from a variety of lipids, and it is probable that they contain vesicles of the same general type. It is probable that any lipid that preferentially adopts the bilayer structure can form such vesicles provided that it contains some unsaturated hydrocarbon chains. There is one report (Saunders, 1966) that they cannot be obtained at room temperature from phosphatidylcholines with identical long saturated fatty acid chains, which would imply resistance to the imposition of curvature on the bilayer because of the tendency to form ordered arrays of the fatty acid chains. More recent work (Sheetz and Chan, 1972), however, has shown that dipalmitoyl phosphatidylcholine does form small vesicles on prolonged sonication, with partial retention of an ordered arrangement of acyl chains.

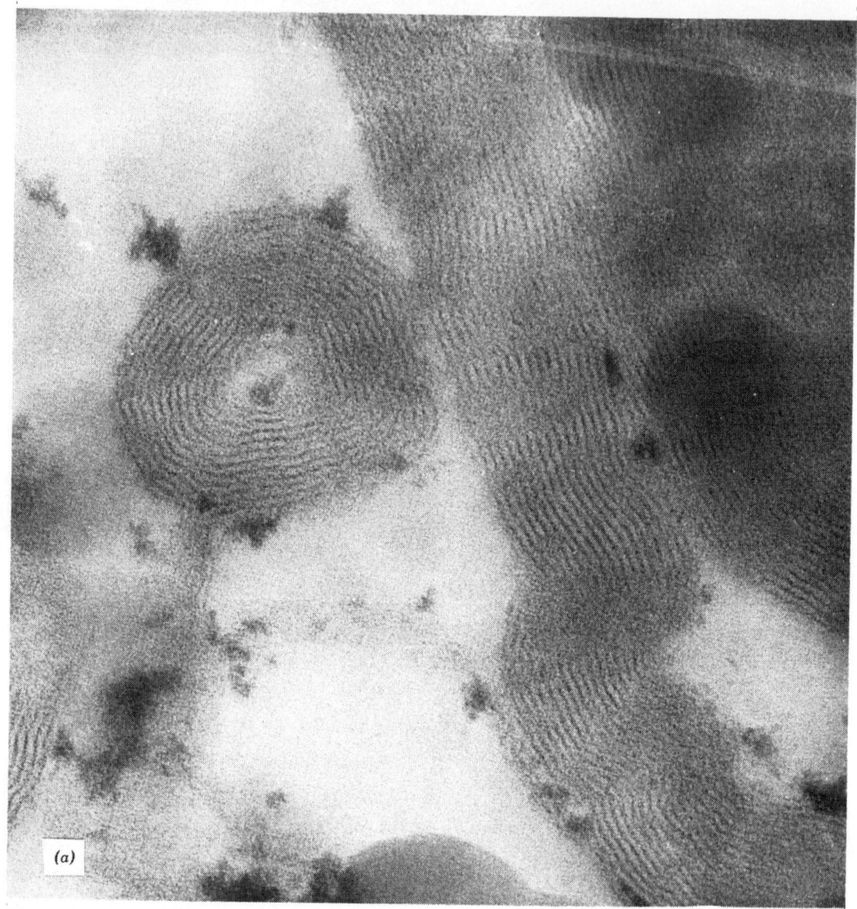

Fig. 12-1. Typical electron micrographs of phospholipid preparations: (*a*) in the form of extended bilayers and multiwalled vesicles (reprinted with permission of Pergamon Press, Ltd., from Dervichian, 1964);

LIQUID CRYSTALLINE PHASES AT HIGH LIPID CONCENTRATIONS

As was true for simple amphiphiles (Chapter 9) biological lipids form a variety of ordered phases (liquid crystals) in mixtures with water when the lipid content is high. As previously noted, these structures tend to persist even when large amounts of water are added, unless disrupted by sonication. In the case of simple amphiphiles the molecular arrangement within the micellar aggregates is greatly altered as the water content is reduced, but for lipids with two hydrocarbon chains per molecule the bilayer arrangement is a common feature of all structures that have been reported, and

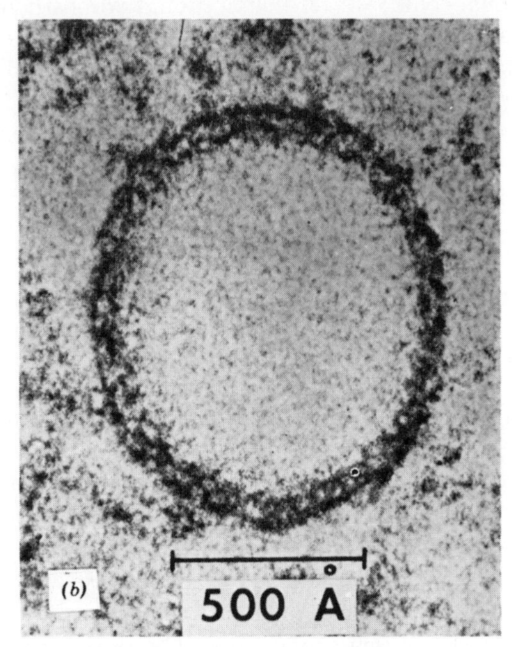

Fig. 12-1. *(continued)* (*b*) in the form of a single-walled vesicle (taken from Miyamoto and Stoeckenius, 1971).

structural studies of the ordered phases are therefore of special interest because they permit determination of bilayer dimensions.

Numerous studies have been carried out by electron microscopy (Fig. 12-1) and X-ray diffraction. The pioneering work by X-ray diffraction was done by Bear et al (1941) and Palmer and Schmitt (1941). They studied a variety of lipids from natural sources, and were able to demonstrate the existence of the bilayer structure and give essentially correct dimensions for it. In particular, they were the first to observe, in anhydrous or nearly anhydrous preparations, the spacing of 4.2 Å characteristic of hexagonally packed hydrocarbon chains perpendicular to the bilayer plane, and to show the broadening of this spacing to an average value of 4.6 Å when water was added. They correctly inferred from this that the interior of the bilayer must then be in a liquidlike state (Schmitt, 1939). More detailed X-ray studies were subsequently made by Luzzati and co-workers, and most of this work has been summarized in a review by Luzzati (1968). Phospholipids with a variety of head groups have been shown to form similar multilamellar structures (Reiss–Husson, 1967; Papahdjopoulos and Miller, 1967), including the phospholipid mixture from mitochondria (Gulik–Krzywicki et al.,

1967), which has a high content of cardiolipin (four hydrocarbon chains per head group). The latter finding is not surprising since the cardiolipin molecule is actually structurally equivalent to two normal phosphoglyceride molecules linked together.

The most accurate bilayer dimensions are probably those given by Levine et al. (1968) and by Levine and Wilkins (1971) who obtained highly ordered stacks of bilayers, separated by water layers, by several techniques. The advantages of such a system are obvious: discrete Bragg reflections are observed in X-ray diffraction instead of continuous diffraction envelopes, and the spacings that they represent are automatically oriented with respect to the direction of the bilayer. Although the overall resolution is relatively poor owing to the rather limited degree of order, a sufficient number of reflections were obtained in the direction perpendicular to the bilayer surface to permit analysis by Fourier synthesis to obtain an electron density profile in that direction. The phase assignment required for this procedure was based on the reasonable assumption that the changes in the repeat distance between bilayers, obtained by swelling the sample with water, are accompanied by relatively small differences in the width of an individual bilayer.

Figure 12-2 represents one repeat of the electron density profile across the stacked bilayers for dry and wet dipalmitoyl phosphatidylcholine, and the interpretation is virtually self-explanatory. The peaks of high electron density must represent the location of phospholipid head groups (which

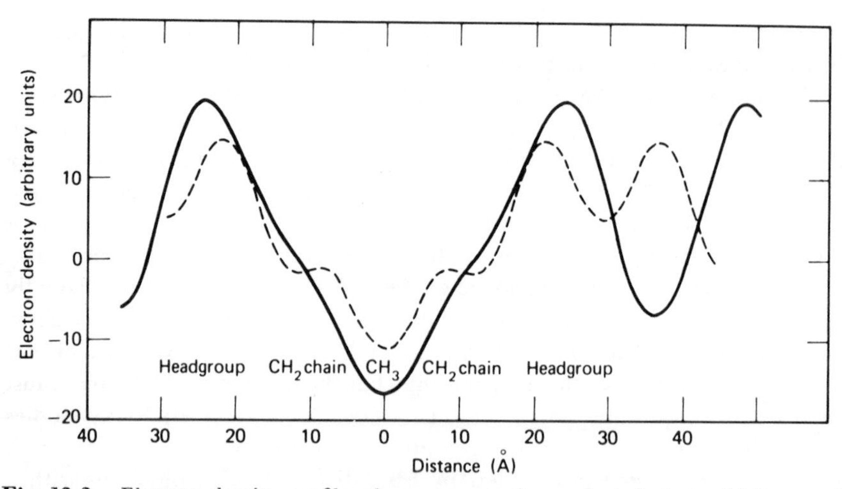

Fig. 12-2. Electron density profile of one repeat of an oriented stack of bilayers of dipalmitoyl phosphatidylcholine (taken from Levine et al., 1968). Electron densities are given relative to liquid water as zero. The dashed line represents lipid in the dry state, the solid line represents data obtained when the system is under water.

contain the electron dense acyl and phosphate ester groups). Two of these peaks are separated by about 16 Å in the dry state, and they must represent head group locations of adjacent bilayers because the separation increases to 25 Å in the wet state. The other two peaks, separated by about 43 Å in the dry state, must represent head group locations on either side of a single bilayer: the increase in their separation in the wet state occurs *outside* of the dry state peaks and represents an extension of the head groups toward the outside, with intercalation of water molecules, and does not involve an increase in the width of the hydrophobic core. The most striking aspect of the profile is the dip in the center of this region, where the electron density falls well below that of water. This dip has been interpreted as representing the location of the methyl groups at the ends of the hydrocarbon chains. Not being covalently linked, the ends of the chains approaching from the two sides would be separated by the relatively large van der Waals distance, which is more than twice the bond distance between covalently linked carbon atoms. (This fact has been previously used in the calculation of micellar volumes in Chapter 9.)

In the direction parallel to the bilayer surface only one reflection at 4.2 Å was observed, corresponding, as previously stated, to the separation between closely packed extended hydrocarbon chains. This distance did not change upon the addition of water, and the overall conclusion is that the bilayer at the temperature of measurement (23°) has a core consisting of extended hydrocarbon chains essentially perpendicular to the bilayer surface, in a regular ordered array.

Similar results were obtained with stacked bilayers consisting of egg yolk phosphatidylcholine, with the important difference that the thickness of the bilayer was found to be smaller and the surface area per head group larger: about 60 $Å^2$ per head group or 30 $Å^2$ per hydrocarbon chain. Comparison with Fig. 9-1 shows that this indicates that the hydrocarbon chains cannot on the average be fully extended in this case, as indeed would not be possible in a mixture containing unsaturated hydrocarbon chains. In addition the 4.2 Å reflection in the bilayer plane was replaced by a diffuse arc at 4.6 Å, showing not only a greater average separation between hydrocarbon chains, but also some variation from the 90° angle relative to the surface. The hydrophobic core in this case is clearly disordered, and the hydrocarbon chains must be undergoing considerable motion even though the bilayers themselves are in fixed positions. However, the trough in electron density at the center of the bilayer is still present, and the change in bilayer dimensions is not large. Substantial orientation of the chains and localization of CH_3 groups near the center evidently persists. It may be noted that increased hydration and separation between adjacent bilayers leads to slightly increased disorientation of the chains, delocalization of the ends, and slight shrinkage of the core thickness. This presumably results from the greater freedom of motion of the head groups.

ARTIFICIAL MEMBRANES FOR PHYSIOLOGICAL STUDIES

For physiological studies (electrochemical properties, diffusion experiments, etc.) artificial membranes of large surface area, consisting of a single lipid bilayer, bounded on both sides by aqueous solutions, have proved to be extremely useful (e.g., Haydon, 1970). Such membranes can be prepared by a procedure first developed by Mueller et al. (1962), which involves extension and drainage of a layer of lipid in an organic solvent, such as decane, until the two monolayers at the water-organic solvent interface coalesce. Very large spherical vesicles with areas up to 1 cm^2 can be formed by closely related procedures (Pagano and Thompson, 1967). The major difference between these systems and the systems already discussed, apart from the larger surface area, is that they retain large amounts of the solvent in which the lipid was originally dissolved (Henn and Thompson, 1968). Thus, although they may serve as suitable models for a phospholipid bilayer from a physiological point of view, they cannot be considered as structurally identical.

CHOLESTEROL AND ITS INCORPORATION IN PHOSPHOLIPID BILAYERS

The state of cholesterol in aqueous solution has not been investigated, undoubtedly because its extremely low solubility makes it difficult to study. Preliminary data (Haberland and Reynolds, 1973) indicate that the monomer concentration at room temperature is limited to $10^{-8}M$, and that a micellelike aggregate exists in solution above that concentration. Because of the rigidity of the sterol ring, this aggregate may not be a micelle of the type discussed in Chapter 6, but rather an aggregate of cholesterol molecules stacked side by side. Formation of such an aggregate would not be expected to be cooperative, that is, the free energy gained by formation of a cholesterol dimer would not be expected to be significantly different from the free energy gained by adding a cholesterol molecule to an already long aggregate. Micelle formation would in that case be a process of gradual growth with increasing concentration, and would not occur as a critical phenomenon within a narrow range of total concentration. In any event the micelles or aggregates themselves have a very limited range of stability and, when the total cholesterol concentration reaches $10^{-6}M$, coalesce and separate from the solution as a separate phase. This process is very likely due to the lack of repulsion between the hydroxyl groups that constitute the sole hydrophilic portion of the cholesterol molecule. We have already noted (page 43) that aliphatic alcohols do not form micelles, but separate as a pure liquid phase instead. Steroid derivatives with charged carboxylate groups (cholanic acids) readily form soluble micelles, although they are often of small size (Small, 1971).

Although pure cholesterol micelles in water present a considerable experimental problem, cholesterol readily enters into micelles formed by phospholipids, and mixed micelles containing as much as equimolar quantities of cholesterol and phospholipid are easily obtained and have received much attention, because the question of the biological significance of the presence of cholesterol in many membranes is an intriguing one. X-ray diffraction studies of oriented bilayers of egg yolk phosphatidylcholine containing an equimolar admixture of cholesterol (Levine and Wilkins, 1971) show that cholesterol increases the average spacing in the direction of the bilayer plane from 4.6 to 4.75 Å. The peak-to-peak distance across the bilayer is significantly increased, and there is considerable sharpening of the electron-density trough at the center of the bilayer, indicative of localization of terminal CH_3 groups over a narrower range. These results are consistent with observations on other systems (Small and Bourgès, 1966; Rand and Luzzati, 1968; Lecuyer and Dervichian, 1970) and suggest that the steroid ring of cholesterol lies in the external portion of the bilayer, as shown in Fig. 12-3,

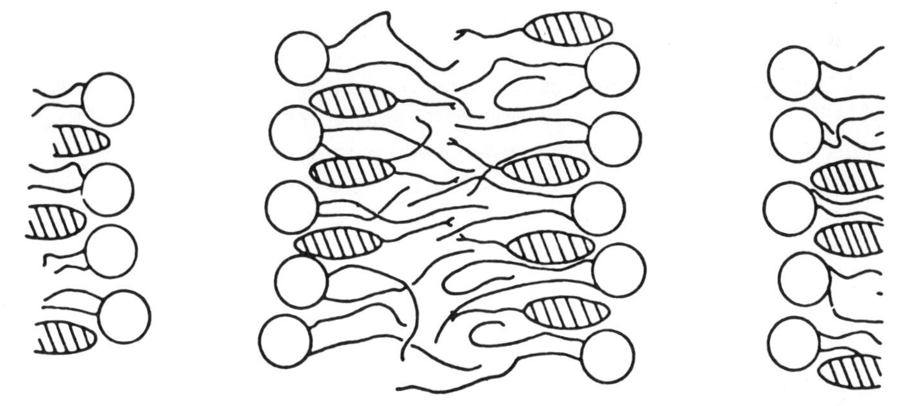

Fig. 12-3. Schematic diagram of the location of cholesterol in phospholipid bilayers (taken from Rand and Luzzati, 1968). The diagram does *not* reflect the subsequent conclusion, discussed in the text, that the phospholipid hydrocarbon chains adjacent to the cholesterol steroid rings (hatched areas) are likely to be in a rigid extended configuration, and in a much more fluid state in the center of the bilayer.

with its hydrophilic OH group in the layer occupied by the phosphatidylcholine head group, whereas the branched aliphatic tail of the molecule (see structural formula at the beginning of this chapter) lies in the central region of the bilayer. The steroid ring is not only rigid, but also has a thicker cross section than the aliphatic tail (Rothman and Engelman, 1972) and the requirement that the total area in the plane of the bilayer be the same at all levels therefore suggests that, in the external part of the bilayer containing the steroid ring, the phospholipid hydrocarbon chains will be fully extended and tend to be very close to perpendicular to the surface. The central part of the bilayer, on the other hand, will need to become very fluid, with many portions of hydrocarbon chains at a considerable angle to the perpendicular direction. The structural detail suggested by this purely geometrical reasoning will be seen to be in accordance with experimental results presented in the following chapter.

EFFECTS OF DETERGENTS

The retention of hydrocarbon in artificial membranes (see above) and the incorporation of cholesterol in phospholipid bilayers are consequences of the nonspecific nature of the hydrophobic force, and represent particular examples of phenomena already considered in a general way in Chapters 6 and 10. Any amphiphilic substance should be able to enter into phospholipid micelles, with the reservation that this could not of course occur for bilayers with an ordered core structure, such as are formed from pure phospholipids with identical saturated acyl chains, without disruption of the ordered structure.

When lysophosphatidylcholine is added to phosphatidylcholine, the bilayer structure is progressively disrupted (Bangham and Horne, 1964), and other simple amphiphiles with a single hydrocarbon chain, such as the common detergents, have the same effect. The phenomenon is again perfectly general and in accordance with the principles of mixed micelle formation set forth in Chapter 10. The equilibrium in a system of mixed amphiphiles is established regardless of the order of addition, and the addition of an excess of detergent or lysophospholipid to a system of phospholipid bilayers must progressively convert the system to one of globular detergent or lysophospholipid micelles in which small amounts of phospholipid with two hydrocarbon chains are incorporated.

It is evident therefore that detergents can be used to disrupt bilayer structures and to disperse biological lipids in soluble form within small detergent micelles. This is not of any special interest here, but it provides a rationale for the use of detergents as a tool in the investigation of biological membranes: as will be shown later (Chapter 19) the lipids of biological membranes are predominantly, though not entirely, in bilayer form. Being inhomogeneous, they may often be selectively disrupted by the addition of detergents.

LIPID MONOLAYERS

Phospholipid monolayers at an air–water or hydrocarbon–water interface resemble monolayers formed by simpler amphiphiles (Chapter 11), with one important difference: because there are two hydrocarbon chains per head group, the head groups cannot approach as close to each other as in monolayers formed by amphiphiles with a single alkyl chain. Thus the minimum area per head group at high pressures is about 40 Å^2 instead of 20 Å^2. In the compression of liquidlike films interactions between head groups are similarly less important than the packing ability of the hydrocarbon chains, as is seen for example from the work of Van Deenen et al. (1962), who obtained quite similar pressure-area curves at the interface between air and 0.14 *M* phosphate buffer for the distearoyl derivatives of phosphatidyl choline, phosphatidyl ethanolamine, phosphatidyl serine and phosphatidic acid, even though the last two have negatively charged head groups. On the other hand, phospholipids containing one unsaturated and one saturated alkyl chain formed considerably more expanded films. The lack of an effect from the repulsion between charged head groups appears to be contradicted by the data of Standish and Pethica (quoted by Pethica, 1969) on films of dipalmitoyl phosphatidylethanolamine. They observed formation of condensed films at relatively low pH, where the head group is zwitterionic, but their result at pH 11.9, where the head group bears a single negative charge, resembles the curve for $C_{18}H_{37}N(CH_3)_3^+$ in Fig. 11-2.

An important observation that has been made in a number of laboratories is that cholesterol does not form liquidlike monolayers at an air–water interface, but condenses at quite low surface pressures to a solidlike film with an area of about 39 Å^2 per molecule (Adam, 1941, p. 49; Pethica, 1969). This result, indicating strong intermolecular forces leading to an ordered aggregate, is consistent with the results obtained for cholesterol aggregates in solution (page 106). It would be of great interest to obtain comparable data at a water-hydrocarbon interface, to determine the extent to which this specific interaction can compete with random solution of the hydrophobic portion of the cholesterol molecule in a hydrocarbon medium, but this experiment has not been done. On the basis of the ease with which cholesterol can be introduced into phospholipid bilayers (in preference to formation of pure cholesterol aggregates) one would anticipate that solidlike monolayers would form at a water-hydrocarbon interface only at high surface pressures.

A number of investigators have reported that cholesterol exerts a condensing effect on phospholipid monolayers, that is, that the area of a mixed phospholipid–cholesterol monolayer at any pressure is smaller than would be predicted on the basis of additivity of the areas of the two components in the mixture (see review by Pethica, 1969). The significance of this observation has been questioned (Gershfeld and Pagano, 1972) because two condensed surface phases appear to coexist in at least some of these experiments.

It may be noted in conclusion that small soluble vesicles bounded by a phospholipid monolayer and containing an inner volume of an organic solvent, can be prepared by sonication of an aqueous dispersion of droplets of lipid in the organic solvent (Träuble and Grell, 1971).

REFERENCES

Abrams, M. E. (1966). *J. Appl. Physiol.*, **21,** 718.
Adam, N. K. (1941). *The Physics and Chemistry of Surfaces*, 3rd. ed., Oxford University Press.
Ames, G. F. (1968). *J. Bact.*, **95,** 833.
Bangham, A. D., and R. W. Horne. (1964). *J. Mol. Biol.*, **8,** 660.
Bangham, A. D., M. M. Standish, and J. C. Watkins. (1965). *J. Mol. Biol.*, **13,** 238.
Bear, R. S., K. J. Palmer, and F. O. Schmitt. (1941). *J. Cell. Comp. Physiol.*, **17,** 355.
DeGier, J., J. G. Mandersloot, and L. L. M. Van Deenen. (1968). *Biochim. Biophys. Acta*, **150,** 666.
Dervichian, D. G. (1964). *Prog. Biophys. and Mol. Biol.*, **14,** 263.
Dickerson, J. W. T. (1968). In *Applied Neurochemistry*, A, N. Davison and J. Dobbing, Eds., F. A. Davis Co., Philadelphia.
Gershfeld, N. L., and R. E. Pagano. (1972). *J. Phys. Chem.*, **76,** 1244.
Gulik-Krzywicki, T., E. Rivas, and V. Luzzati. (1967). *J. Mol. Biol.* ,**27,** 303.
Haberland, M. E., and J. A. Reynolds. (1973). *Fed. Proc.*, **32,** 639 Abs.
Haydon, D. A. (1970). In *Membranes and Ion Transport*, E. E. Bittar, Ed., Vol. 1, p. 64. John Wiley and Sons, New York.
Henn, F. A., and T. E. Thompson. (1968). *J. Mol. Biol.*, **31,** 227.
Hill, E. E., and W. E. M. Lands. (1970). In *Lipid Metabolism*, S. Wakil. Ed., Academic Press, New York, Chapter 6.
Houtsmuller, U. M. T., and L. L. M. Van Deenen. (1965). *Biochim. Biophys. Acta*, **106,** 564.
Huang, C. (1969). *Biochemistry*, **8,** 344.
Huang, C., L. Wheeldon, and T. E. Thompson. (1964). *J. Mol. Biol.*, **8,** 148.
Lecuyer, H., and D. G. Dervichian. (1969). *J. Mol. Biol.*, **45,** 39.
Lennarz, W. J. (1972). *Acc. Chem. Res.*, **5,** 361.
Levine, Y. K., and M. H. F. Wilkins. (1971). *Nature New Biol.*, **230,** 69.
Levine, Y. K., A. I. Bailey, and M. H. F. Wilkins. (1968). *Nature*, **220,** 577.
Lewis, M. S., and M. H. Gottlieb. (1971). *Fed. Proc.*, **30,** 1303 Abs.
Luzzati, V. (1968). In *Biological Membranes*, D. Chapman, Ed., Academic Press, New York, Chapter 3.

McElhaney, R. N., and M. E. Tourtelotte. (1969). *Science*, **164**, 433.
Meldolesi, J., J. D. Jamieson, and G. E. Palade. (1970). *J. Cell Biol.*, **49**, 130.
Miyamoto, V. K., and W. Stoeckenius. (1971). *J. Membrane Biol.*, **4**, 252.
Mueller, P., D. O. Rudin, H. T. Tien, and W. C. Westcott. (1962). *Nature*, **194**, 979.
Pagano, R., and T. E. Thompson. (1967). *Biochim. Biophys. Acta*, **144**, 666.
Palmer, K. J., and F. O. Schmitt. (1941). *J. Cell. Comp. Physiol.*, **17**, 385.
Papahdjopoulos, D., and N. Miller. (1967). *Biochim. Biophys. Acta*, **135**, 624.
Pethica, B. A. (1969). In *Structural and Functional Aspects of Lipoproteins in Living Systems*, E. Tria and A. M. Scanu, Eds., Academic Press, New York.
Rand, R. P., and V. Luzzati. (1968). *Biophys. J.*, **8**, 125.
Reiss-Husson, F. (1967). *J. Mol. Biol.*, **25**, 363.
Robinson, N., and L. Saunders. (1958). *J. Pharm. Pharmacol.*, **10**, 755.
Rothman, J. E., and D. M. Engelman (1972). *Nature New Biol.*, **237**, 42.
Rouser, G., G. J. Nelson, S. Fleischer, and G. Simon. (1968). In *Biological Membranes*, D. Chapman, Ed., Academic Press, New York, Chapter 2.
Saunders, L. (1966). *Biochim. Biophys. Acta*, **125**, 70.
Saunders, L., J. Perrin, and D. Gammack. (1962). *J. Pharm. Pharmacol.*, **14**, 567.
Schmitt, F. O. (1939). *Physiol. Revs.*, **19**, 270.
Sheetz, M. P., and S. I. Chan. (1972). *Biochemistry*, **11**, 4573.
Small, D. M. (1971). In *The Bile Acids*, P. P. Nair and D. Kritchevsky, Eds., Plenum Press, New York, Chapter 8.
Small, D. M., and M. Bourgès. (1966). *Mol. Cryst.*, **1**, 541.
Smith, R., and C. Tanford. (1972). *J. Mol. Biol.*, **67**, 75.
Träuble, H., and E. Grell. (1971). *Neurosci. Res. Prog. Bull.*, **9**, 373.
Van Deenen, L. L. M., U. M. T. Houtsmuller, G. H. De Haas, and E. Mulder. (1962). *J. Pharm. Pharmacol.*, **14**, 429.
Winkler, H., and A. D. Smith. (1968). *Arch. Pharmak. Exp. Path.*, **261**, 379.

THIRTEEN

MOTILITY AND ORDER

It was noted in Chapter 6 that long-chain unbranched saturated hydrocarbons form crystalline phases with melting points near or above room temperature, whereas the melting points of unsaturated hydrocarbons are much lower. X-ray diffraction data presented in the preceding chapter indicated that a somewhat analogous situation exists in the hydrophobic core of bilayer structures formed by biological lipids containing two hydrocarbon chains per molecule: in a pure lipid with two saturated hydrocarbon chains the chains appear to be in an ordered array at room temperature, whereas they are disordered in bilayers formed by lipid mixtures containing both saturated and unsaturated hydrocarbon chains. In the present chapter additional information obtainable by other methods will be examined to refine our picture of the state of the bilayer core. In addition, measurements of the motility of individual lipid molecules in and out of the bilayer and within it will be discussed.

FLUIDITY OF THE BILAYER INTERIOR

Pure anhydrous phospholipids are solids with very high melting points. The melting points are unrelated to hydrocarbon chain length or degree of unsaturation. This in itself suggests that the hydrocarbon chains are already in a liquidlike state in the solid at the melting point, and that the melting process primarily reflects the disruption of intermolecular bonds between head groups. Support for this conclusion is provided by the observation of a major phase transition within the solid state at temperatures below the true melting point. This phase transition occurs at the same temperature as the transition of the interchain X-ray spacing from 4.2 to 4.6 Å in those instances where X-ray diffraction measurements as a function of temperature have been made, and is generally taken to reflect an order–disorder transition of the hydrocarbon chains within the solid. Similar phase transitions, but at temperatures below those at which they occur in the anhydrous solid, can be observed in liquid crystalline phases formed by mixtures of phospholipids with water and in sonicated dispersions.

A direct (but mechanistically uninformative) way to observe the transition is by differential thermal analysis (see review by Ladbrooke and Chapman, 1969). In this method a profile of heat absorption versus temperature is obtained. In the absence of phase transitions a monotonic line is obtained, the slope of which measures the heat capacity of the system. Phase transitions are characterized by additional heat absorption over narrow temperature ranges: both the transition temperature and the enthalpy change in the process are directly determinable. Examples of data obtained by this method are given in Table 13-1. They show at once that the observed process is not directly comparable to the melting of hydrocarbon crystals, for which data were given in Table 6-2. The observed transition temperatures are higher

Table 13-1. Thermodynamics of Transition from Ordered to Disordered Hydrocarbon Chains in Diacyl Phosphatidylcholines[a]

	Anhydrous Crystalline State		Hydrated Liquid Crystalline State[d]	
Acyl Groups[b]	Transition Temperature (°C)	ΔH^c (kcal/mole)	Transition Temperature (°C)	ΔH^c (kcal/mole)
C_{22}, saturated	120	8.4	75	7.4
C_{18}, saturated	115	6.9	58	5.3
C_{16}, saturated			41	4.3
C_{14}, saturated			23	3.3
C_{18}, monounsaturated			−22	3.8

[a] From Phillips et al. (1969).
[b] Each molecule contained two identical chains. The numerical subscript indicates the number of carbon atoms of the acyl chain, including that of the ester group. The number of carbon atoms in the hydrocarbon chain is one less than this.
[c] To facilitate comparison with Table 6-2, ΔH values are given per hydrocarbon chain, that is, they are half the values per mole of phospholipid.
[d] The data represent the liquid crystalline phases with maximal water content, that is, similar to those to which the solid line of the X-ray data of Fig. 12-2 are applicable.

than the melting points of hydrocarbons of comparable chain length. They are much higher in anhydrous crystalline phospholipids than in hydrated liquid crystals, and this implies that the transition involves the phospholipid head groups as well as the hydrocarbon tails. This feature of the transition is also indicated by considerable differences in the transition temperature between phospholipids with different head groups: for example, the transition temperature for dipalmitoyl phosphatidylethanolamine is about 20° higher than that for dipalmitoyl phosphatidylcholine under comparable experimental conditions (Abramson, 1969). Direct evidence that a loosening of head group organization accompanies the order–disorder transition of the hydrocarbon chains has been given by Träuble (1971). He has also shown that the transition is highly cooperative, as is to be expected.

Table 13-1 also shows that the ΔH values are lower than the ΔH values for the melting of pure hydrocarbons. The same is true for the entropy change in the process (Phillips et al., 1969) which is only about half as large as the ΔS values given in Table 6-2 for hydrocarbons with an odd number of carbon atoms, and even smaller in relation to hydrocarbons with an even number of carbon atoms. This must mean that the hydrocarbon chains in the "melted" bilayer core are not as disordered as they are in a pure liquid hydrocarbon. Nevertheless, the effects of hydrocarbon chain length and of unsaturation are qualitatively similar for the two processes. All the results are consistent with the idea already put forward in Chapter 6 that the restrictions of hydrocarbon chains in micellar aggregates are likely to be greater near the point of attachment to hydrophilic groups than at the free ends of the chains, and this will be demonstrated conclusively below.

The transition from ordered to disordered hydrophobic cores can also be observed by spectroscopic methods (Chapman and Wallach, 1968; Träuble, 1971). Such studies strongly support the inferences from X-ray diffraction and differential thermal analysis and, in particular, leave no doubt about the liquidlike nature of the bilayer interior above the transition temperature. This is evident, for example, from the high resolution pmr spectrum of dispersions of egg yolk phosphatidylcholine (Chapman and Penkett, 1966). The transition temperature of these mixed bilayers is below 0°C, as previously noted. At room temperature the pmr absorption peaks corresponding to hydrocarbon CH_2 groups, as well as those from the terminal CH_3 groups of the hydrocarbon chains and from the head group $N(CH_3)_3^+$ protons, are relatively sharp, indicating sufficiently rapid motion to permit time-averaging of different orientations with respect to the magnetic field in 10^{-8} sec or less (page 41). The observed spectra are quite similar to those observed for true solutions of phospholipids in an organic solvent such as $CHCl_3$.

MORE DETAILED STUDY OF HYDROCARBON CHAIN MOTILITY

More detailed pmr measurements (Chan et al., 1971) indicate that not all of the CH_2 groups are capable of rapid motion, whereas all of the CH_3 groups are, indicating, as previously suggested, that internal motion is restricted as the bilayer surface is approached. A similar result, with far better resolution, has since been obtained by use of the nmr spectrum of the ^{13}C isotope of carbon (Levine et al., 1972). In principle this method permits independent observation of each individual C atom in the structure by specific enrichment of the ^{13}C isotope at each position.

Another way to obtain direct information on this question is by use of paramagnetic free radical probes of the type

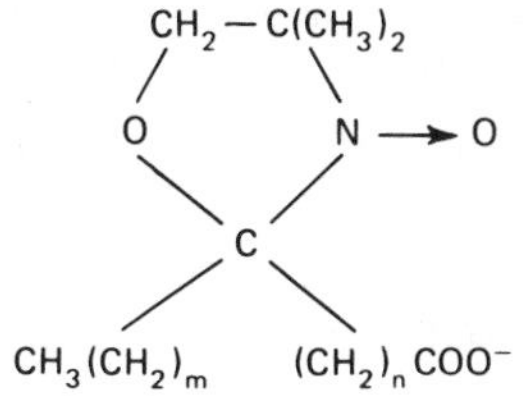

$$\begin{array}{c} CH_2 - C(CH_3)_2 \\ / \qquad \quad \backslash \\ O \qquad \qquad N \rightarrow O \\ \backslash \qquad / \\ C \\ / \qquad \backslash \\ CH_3(CH_2)_m \qquad (CH_2)_n COO^- \end{array} \qquad (I)$$

These probes orient themselves in a bilayer with the COO^- group at the hydrophilic surface, with the labeled hydrocarbon chain preferentially parallel to the hydrocarbon chains of the lipid molecules. By varying m and n in formula I different depths of the bilayer can be probed, using the interpretive principles briefly discussed in Chapter 6.

Hubbell and McConnell (1968, 1971) have measured the esr spectrum of probes of type I in dispersions of egg yolk phosphatidylcholines and mixed lipid preparations from soy beans. With $n = 3$, that is, with the $N \rightarrow 0$ group close to the bilayer surface, both parallel and perpendicular components of the hyperfine splitting could be observed, showing that segmental motion of the hydrocarbon chains at that location must be highly anisotropic. There was rapid rotational motion about an axis perpendicular to the bilayer surface, but much less motion in the other direction: the mean angular deviation from a strictly perpendicular orientation of the hydrocarbon chain was only 26°. The same result was obtained with $m = 12$ and $m = 17$ in formula I, that is, the overall length of the hydrocarbon chain had no effect when the separation between the spin label of the probe molecule and the bilayer surface remained unchanged. On the other hand, when this separation was increased ($n = 10$ in formula I) distinct parallel and perpendicular components of the hyperfine splitting could no longer be observed, showing that the molecular motion was experimentally indistinguishable from that in an isotropic liquid medium. The fluidity in the center of the bilayer is similar to that of the interior of globular micelles formed by simple amphiphiles, as discussed in Chapter 6. Essentially the same result, using similar procedures, has been obtained by Jost et al. (1971).

A problem associated with the increase in disorder as one passes from the exterior to the interior of the bilayer is that freely rotating segments of hydrocarbon chains occupy greater volume per unit mass than do ordered parallel chains, that is, the density of liquid hydrocarbon is less than that of crystalline solid. This means that an extended bilayer cannot have *parallel* arrays of ordered hydrocarbon chains in one region and liquidlike chains in another because this would leave free space in the ordered region. McFarland and McConnell (1971) have suggested that the simplest way to overcome this difficulty is to tilt the hydrocarbon chains in the ordered region, whereby the thickness of that region would be decreased, thus eliminating the free space, as shown schematically in Fig. 13-1. They have provided evidence for a structure of this type by repeating the spin label experiments described above on an *ordered* multilayer of egg phosphatidylcholine. They used probes such as described by formula I as well as nitroxide derivatives of diacyl phosphatidylcholine in which the label was placed in various positions on one of the hydrocarbon chains. Because the orientation of the sample with respect to the magnetic field is fixed in this experiment, the mean angular deviation of the external parts of the paraffin chains from strictly perpendicular, as previously observed with phospholipid suspensions, could be decomposed into a net average tilt and a fluctuating component. A net average tilt of 25 to 30° was found to represent the major part of the mean angular deviation. When the label was attached at a distal point of the hydrocarbon chain, the net average tilt became close to zero, as expected.

Similar information should in principle be obtainable by use of fluorescent probes (Waggoner and Stryer, 1970), but directly comparable studies by this procedure have not been reported.

There is of course a danger in placing too much reliance on experiments utilizing paramagnetic or fluorescent probes, since they may significantly perturb the bilayer structure (Leslauer et al., 1972). The general consistency between the results obtained by use of nitroxide spin labels and those ob-

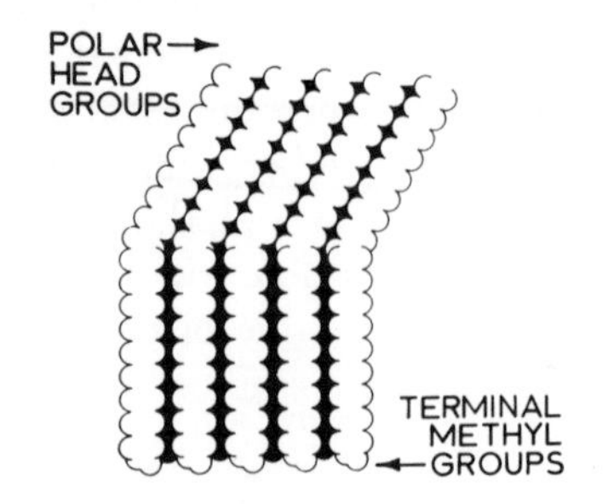

Fig. 13-1. Schematic representation of the packing of hydrocarbon chains in a small section of one-half of a planar bilayer of diacyl phosphatidylcholine in the fluid state (from McFarland and McConnell, 1971). The unfilled space between chains in the central part of the bilayer provides freedom for random liquidlike motion of the chains.

tained from X-ray diffraction or pmr measurements suggests that the perturbations are not serious in this case. Another piece of evidence is provided by the work of Hubbell and McConnell (1971) with dispersions of dipalmitoyl phosphatidylcholine containing a small proportion of nitroxide-labeled molecules. The order–disorder transition was observed to occur at essentially the same temperature as in the absence of labeled molecules. Thus, although some perturbation of the packing in the vicinity of the labeled molecules seems inevitable, the necessary interplay between the labeled chain and adjacent unlabeled chains evidently minimizes the effect of the perturbation.

It is useful at this point to return to globular micelles of simple amphiphiles such as were discussed in Chapters 6 and 9. Seelig (1970) has carried out experiments with paramagnetic probes of type I on liquid crystals containing alkyl carboxylates, at low water content, where the bilayer is the principal structural element (see page 74). His results were similar to those obtained for phospholipid bilayers, that is, the fatty acid hydrocarbon chains are in an ordered close-packed array near the bilayer surface and liquidlike near the center. It would seem highly probable that the spatial requirements of a bilayer structure per se (as in Fig. 13-1) are crucial to this result, and that it is not reasonable to extend this finding to *globular* micelles formed by alkyl carboxylates or other simple ionic amphiphiles, if only for the simple geometrical reason that the head groups of a globular micelle lie on a highly curved surface and are relatively far apart. Hydrocarbon chains near the surface of a globular micelle may be constrained because they are anchored to the head group, but the constraint is such as to oppose rather than favor a close-packed arrangement.

EFFECT OF CHOLESTEROL

Results of investigations of the effect of cholesterol on phospholipid bilayers have been summarized by Rothman and Engelman (1972). They are consistent with the location of cholesterol in the bilayer shown in Fig. 12-3. The effect on the first six or seven carbon atoms of the phospholipid chain, which are in contact with the relatively bulky steroid ring, is to *decrease* motility, and, presumably, to decrease the angle of tilt (Fig. 13-1), although actual data on this point have not been obtained. The effect on the central portion of the bilayer, where the branched paraffin chain of the cholesterol molecule is located, is to *increase* fluidity.

MOBILITY OF LIPID MOLECULES

The ability to define and measure a unique monomer concentration in equilibrium with phospholipid dispersions implies the existence of a dynamic equilibrium between lipid molecules in a bilayer and in the solution with which it is in contact. However, the fact that the monomer concentration is very small, of order $10^{-10}M$ for typical lipids (Table 12-2), means that the rate of exchange between bilayer and surroundings must be slow. No direct measurements have been made, but Kornberg and McConnell (1971a) have measured the rate of exchange of spin–labeled phosphatidylcholine between small phosphatidylcholine vesicles which, under their experimental conditions, was very slow indeed (average exchange time of the order of 24 hr). It is not known whether exchange occurred in their experiments by direct transfer between colliding vesicles, or by dissociation of a labeled molecule from one vesicle, followed by diffusion and incorporation into a new vesicle.

Another slow process is the interchange of lipid molecules between one side of the bilayer and the other. This process requires passage of the hydrophilic head group of the lipid through the hydrophobic core, and is therefore associated with a very high free energy of activation. Measurements of the rate of the process have been made by Kornberg and McConnell (1971b). They used a phosphatidylcholine with a nitroxide spin label attached to the choline head group and made use of the fact that the nitroxide group is readily reduced to an inert product by ascorbic acid. Egg phosphatidylcholine vesicles similar to those described on page 101 were prepared, bounded by a single bilayer, and containing an internal solvent-filled cavity. A small percentage of spin-labeled molecules was added to the lipid from which the vesicles were made, and was found to distribute itself more or less randomly between the outer and inner surface of the bilayer: 65% of the initial esr intensity could be removed by reduction with ascorbate, which had been shown previously not to enter the internal cavity of the vesicle. The rate of interchange between the two bilayer surfaces was then measured by following the appearance of more reducible label as a function of time: this was taken to represent labeled molecules that had crossed from the inner to the outer surface. The half-time of the process was found to be 6.5 hr at 30°, corresponding to a probability of 0.07/hr for the passage of a given molecule from the internal to the external surface.

In contrast to the two preceding kinds of motion, lateral diffusion of phospholipid molecules in the plane of the bilayer is extremely rapid, so that lipid molecules with labeled head groups introduced at high concentration at one point of a bilayer vesicle, for example, would become uniformly distributed over the entire surface in less than a second. The diffusion coefficient has been evaluated by Devaux and McConnell (1972), again by use of nitroxidelabeled phosphatidylcholine. The broadening of the resonance line in the esr spectrum by spin–spin interaction with increasing concentration of the label was used to establish an empirical measure of concentration. Ordered multilayers (page 104) of lipid containing a high concentration of labeled molecules were then laid down[1] and fused with multilayers containing no label. The rate of diffusion of the label into the unlabeled areas was then measured. The diffusion coefficient was found to be $1.8 \pm 0.6 \times 10^{-8}$ cm^2/sec at 25°, corresponding to an average displacement of about 10^{-4} cm per second as a result of the random motion of a single molecule.

Kornberg and McConnell (1971a) obtained a qualitatively consistent result for egg phosphatidylcholine vesicles by a different procedure. Träuble and Sackman (1972), also using a somewhat different procedure, obtained a diffusion coefficient of about 10^{-8} cm^2/sec for a spin-labeled steroid in a dipalmitoyl phosphatidylcholine *monolayer* vesicle (page 110) when measurements were made at temperatures above the transition point for the order-disorder transition of this phospholipid. Similar experiments below the transition point indicated that rapid motion in the plane of the monolayer does not take place when the lipid molecules are in an ordered close-packed array.

REFERENCES

Abramson, M. A. (1969). In *Surface Chemistry of Biological Systems*, M. Blank, Ed., Plenum Press, New York, p. 37.

Chan, S. I., G. W. Feigenson, and C. H. A. Seiter. (1971). *Nature*, **231**, 110.

Chapman, D., and D. F. H. Wallach. (1968). In *Biological Membranes*, D. Chapman, Ed., Academic Press, Chapter 4.

Chapman, D., and S. A. Penkett. (1966). *Nature*, **211**, 1304.

Devaux, P., and H. M. McConnell. (1972). *J. Am. Chem. Soc.*, **94**, 4475.

Hubbell, W. L., and H. M. McConnell. (1968). *Proc. Nat. Acad. Sci. U.S.A.*, **61**, 12.

Hubbell, W. L., and H. M. McConnell. (1971). *J. Am. Chem. Soc.*, **93**, 314.

Jost, P., L. J., Libertini, V. C. Herbert, and O. H. Griffith. (1971). *J. Mol. Biol.*, **59**, 77.

Kornberg, R. D., and H. M. McConnell. (1971a). *Proc. Nat. Acad. Sci. U.S.A.*, **68**, 2564.

Kornberg, R. D., and H. M. McConnell. (1971b). *Biochemistry*, **10**, 1111.

Ladbrooke, B. D., and D. Chapman. (1969). *Chem. Phys. Lipids*, **3**, 304.

Leslauer, W., J. E. Cain, and J. K. Blasie. (1972). *Proc. Nat. Acad. Sci. U.S.A.*, **69**, 1499.

Levine, Y. K., N. J. M. Birdsall, A. G. Lee, and J. C. Metcalfe. (1972). *Biochemistry*, **11**, 1416.

McFarland, B. G., and H. M. McConnell. (1971). *Proc. Nat. Acad. Sci. U.S.A.*, **68**, 1274.

Phillips, M. C., R. M. Williams, and D. Chapman. (1969). *Chem. Phys. Lipids*, **3**, 234.

Rothman, J. E., and D. M. Engelman. (1972). *Nature New Biol.*, **237**, 42.

Seelig, J. (1970). *J. Am. Chem. Soc.*, **92**, 3881.

Träuble, H. (1971). *Naturwissenschaften*, **58**, 277.

Träuble, H., and E. Sackman. (1972). *J. Am. Chem. Soc.*, **94**, 4499.

Waggoner, A. S., and L. Stryer. (1970). *Proc. Nat. Acad. Sci. U.S.A.*, **67**, 579.

[1] The bilayers in most of these experiments were prepared from dihydrosterculoyl phosphatidylcholine, a synthetic lipid analogous to dioleoyl phosphatidylcholine, but containing a cyclopropane ring in place of the double bond. This was done because the high concentration of the free radical label required for this experiment leads to slow peroxidation of double bonds when they are present. The hydrosterculoyl chain is sterically very similar to the oleoyl chain, as the cyclopropane ring introduces a fixed 120° bond angle into the structure just as a double bond does.

FOURTEEN

PROTEINS

HYDROPHOBIC SIDE CHAINS AND CONFORMATIONAL CHANGE

Proteins are amphiphilic molecules. They contain amino acids with hydrocarbon and other hydrophobic side chains, and also amino acids with ionic and uncharged polar side chains. The polypeptide backbone itself contains the strongly hydrophilic peptide group, but its effect can be nullified by formation of interpeptide hydrogen bonds, such as exist in the α-helical and parallel extended chain β-structures of the polypeptide chain (Pauling et al., 1951). The network of hydrogen bonds in such structures appears to have comparable stability with hydrogen bonds between the peptide groups and water, and polypeptide *backbones* that have adopted this type of conformation no longer retain a marked preference for being in an aqueous environment.

Measurements of the solubilities of amino acids and related substances, in water and in organic solvents such as ethanol, have been used to assess the contributions of the side chains to the free energy of transfer from water to the organic solvent (Cohn and Edsall, 1943; Nozaki and Tanford, 1971). The numerical values of these parameters are shown in Table 14-1 and are seen to be considerably smaller than the values of $\mathring{\mu}_{HC} - \mathring{\mu}_{W}$ for pure hydrocarbons given in Chapter 2. For example, taking the difference between $\mathring{\mu}_{\mathrm{EtOH}} - \mathring{\mu}_{W}$ for norleucine and glycine as representing the contribution of the moiety $—(CH_2)_3—CH_3$ one obtains $-$ 2600 cal/mole at 25°C. The contribution of the same group to $\mathring{\mu}_{HC} - \mathring{\mu}_{W}$ of alkanes (equation 2-4) would be -4750 cal/mole. Part of the reason for this discrepancy arises from the fact that ethanol is not directly comparable to pure hydrocarbon as a solvent, because it must possess a solvent "structure" that will be perturbed by addi-

Table 14-1. Hydrophobicity of Protein Side Chains[a, b]

Side Chain	$\overset{\circ}{\mu}_{org} - \overset{\circ}{\mu}_{W}$ (cal/mole)
Tryptophan	−3400
Norleucine[c]	−2600
Phenylalanine	−2500
Tyrosine	−2300
Leucine	−1800
Valine	−1500
Methionine	−1300
Alanine	− 500

[a] The figures given represent the *additional* free energy of transfer from water to an organic solvent that is generated when the side chain listed is substituted for the hydrogen atom of a glycyl residue. Most of the data are derived from experiments in which the organic solvent was ethanol or dioxane at 25°C (from Nozaki and Tanford, 1971).

[b] The principal hydrophobic side chains include, in addition to those listed, isoleucine, proline, and cystine. Experimental values of $\overset{\circ}{\mu}_{org} - \overset{\circ}{\mu}_{W}$ are not available for them. Portions of some hydrophilic side chains are hydrophobic, notably the $—(CH_2)_4—$ moiety of lysine, which is likely to be characterized by a $\overset{\circ}{\mu}_{org} - \overset{\circ}{\mu}_{W}$ value of −1.0 to −1.5 kcal/mole. This free energy decrease can of course be realized only if the terminal $-NH_3^+$ group remains in contact with water at the same time as the $—(CH_2)_4—$ moiety is removed from such contact.

[c] Norleucine is not a naturally occurring side chain and is included here as a reference compound. Essentially identical results have been obtained with ethanol, butanol, and acetone as the organic solvent.

tion of a solute. Table 2-2 suggests that 500 to 1000 cal/mole of the difference may be ascribable to this factor. A second contributing factor is the short length of the hydrocarbon moiety. As previously stated, that part of a hydrocarbon chain that is at or near the point of attachment to a polar group is not as hydrophobic as a similar portion of a pure hydrocarbon molecule or a portion of hydrocarbon on an amphiphile molecule that is far removed from the polar head group. Despite these quantitative differences there is no question that the hydrocarbon side chains are hydrophobic, and that the magnitude of the unfavorable free energy of forming an interface with water increases with the size of the side chain: the aromatic side chains are larger than aliphatic ones and make the largest contributions.

In water-soluble proteins about 25 to 30% of the amino acid side chains are generally significantly hydrophobic, and 45 to 50% are typically ionic or contain uncharged hydrophilic side chains. The rest (in which we have included alanine from among the side chains listed in Table 14-1) have relatively little preference for being in or out of the aqueous environment. In the native conformation of such a protein, a substantial fraction of the hydrophobic side chains are typically buried in the interior of the molecule, the free energy gained thereby being a major factor in the stability of the native conformation (in water) relative to a more flexible conformation in which the hydrophobic side chains would be exposed to solvent (Tanford, 1962). The native conformation is, in fact, in dynamic equilibrium with other possible conformations, and reversible transitions to other conformations take place when the solvent medium is altered (Tanford, 1968). The transition from a disordered state to the native conformation thus has some resemblance to micelle formation, and, like micelle formation, it is a highly cooperative process: structures in which just a few hydrophobic side chains are removed from an aqueous environment are generally not stable.

The similarity between micelle formation and protein folding is, however, only superficial. The hydrophobic amino acids are not grouped together in the polypeptide chain, but are interspersed with amino acids having ionic or polar side chains, each individual protein polypeptide chain having a unique sequence. Moreover, because the side chains are so short, much of the polypeptide backbone to which they are attached has to be incorporated into hydrophobic domains that are formed in the interior of the native structure, and, as noted above, this requires the formation of a network of interpeptide hydrogen bonds in order to nullify the otherwise strong affinity of the peptide group for water. Polar uncharged side chains are under a similar constraint, and, if located in a portion of the polypeptide chain that is to be incorporated in a hydrophobic domain, must form hydrogen bonds to other polar groups if the free energy is to be minimized.

The most severe restriction of all applies to the ionic amino acid side chains of arginine, lysine, glutamic acid, and aspartic acid. These side chains cannot be removed from an aqueous environment without severe loss of free energy. As in the formation of ionic micelles, they must remain at the surface of the globular structure, and, indeed, virtually no exceptions to this principle have been found in proteins whose three-dimensional structure has been determined. Since the ionic side chains cannot be physically separated from neighboring hydrophobic amino acids, the limitations on possible stable structures are obvious, and polypeptide chains containing the normal proportions of the various kinds of amino acids, in an arbitrary sequence, may not be able to form stable globular structures at all. The amino acid sequences found in nature may represent a special group, selected by the evolutionary process because they are capable of forming structures in which hydrophobic amino acids can be segregated in internal domains without violating the requirements regarding location of ionic side chains and hydrogen bonding between peptide groups and other internal polar groups.

One of the consequences of the foregoing restrictions is that the structure of native proteins must be quite rigid. The internal hydrophobic domains must be in a uniquely ordered state, quite different from the fluid interior of a micelle. As a result of this, the free energy gained by removing hydrophobic side chains from water in formation of the native structure is actually much less than the free energies of transfer listed in Table 14-1 would predict. It is, in fact, found experimentally that native structures frequently have only marginal stability relative to disordered conformations or alternate structures (see below).

BINDING SITES FOR HYDROPHOBIC LIGANDS

Another consequence of the difficulty in the formation of a stable globular structure is that *complete* removal of hydrophobic side chains from contact with water is generally not possible. In most native proteins some hydrophobic groups remain exposed at the molecular surface or in crevices. If sufficiently large hydrophobic patches are formed, they may constitute binding sites for hydrocarbon or amphiphile molecules. The standard free energy of transfer of ligand molecules to such sites on the protein molecule ($\mu_P^\circ - \mu_W^\circ$) may be more negative than $\mu_{HC}^\circ - \mu_W^\circ$ for a hydrocarbon or $\mu_{\text{mic}}^\circ - \mu_W^\circ$ for an amphiphile because the process of binding may involve elimination of a hydrocarbon-water interface at the protein surface as well as that associated with the free ligand in the aqueous medium. Other factors exist, of course, that may diminish the affinity between protein and hydrophobic ligand. For example, the hydrocarbon-water interface may be only partially eliminated by binding to a surface site, or the bound ligand may be required to adopt an unfavorable conformation when it is bound. Another possibility is that energetically unfavorable local rearrangements of protein structure near the binding site may occur as part of the association process.

In the association of amphiphiles with hydrophobic binding sites of native proteins, the hydrophilic head group may play a significant role. An obvious example would be an ionic amphiphile. The electrostatic repulsion between head groups makes a significant positive contribution to $\mu_{\text{mic}}^\circ - \mu_W^\circ$, electrostatic attraction between a head group and an ionic protein side chain with opposite charge could make a negative contribution to $\overset{\circ}{\mu}_P - \overset{\circ}{\mu}_W$.

The overall conclusion from these considerations is that native protein molecules may have hydrophobic sites suitable for association with hydrocarbons or amphiphiles, and that the free energy associated with binding to these sites is in general unpredictable and may be larger or smaller than the free energy associated with competing processes such as micelle formation.

CONFORMATIONAL CHANGE

The native conformation of a protein molecule possesses only marginal stability because it is highly constrained. Other conformational states, in which a much larger fraction of the hydrophobic side chains is exposed to the solvent than in the native state, are thus readily accessible (Tanford, 1968). One such state is the randomly coiled state, in which the protein polypeptide chains become flexible and all hydrophobic residues are exposed to the solvent. The free energy of transition to this state is small even in aqueous solution, typically about 100 cal/mole per amino acid residue (Tanford, 1970), because the creation of a large hydrocarbon-water interface is largely compensated for by increased freedom of internal motion of the polypeptide chain.

Another denatured state which, like the randomly coiled state, may be accessible to most proteins, is characterized by having a large fraction of the polypeptide chains in an α-helical conformation. It has never been carefully investigated, but it is likely to consist of several lengths of helical polypeptide chain, with flexible joints between them, and most of the hydrophobic side chains exposed to solvent. This state is commonly observed when proteins are dissolved in mixtures of water with alcohols and other organic solvents, in which exposure of hydrophobic residues is no longer attended by an unfavorable free energy change. The free energy of transition in a purely aqueous solution has not been determined, but it is likely to be higher than the free energy of transition to the randomly coiled state.

Protein molecules in such altered conformational states, with many exposed hydrophobic groups, are likely to be able to bind a large number of hydrophobic ligands per molecule. The free energy gained thereby (with $\overset{\circ}{\mu}_P - \overset{\circ}{\mu}_W$ intrinsically very negative) may greatly exceed the unfavorable free energy change accompanying the conformational change per se, and the transition to the altered conformational change would then be induced by the presence of the ligand. It is actually well known that detergents are good denaturing agents, and that they induce formation of a "helical" state that may be similar to the state attained by proteins in mixtures of water with organic solvents (Tanford, 1968; Jirgensons and Capetillo, 1970).

The formation and gross disruption of the native conformation of proteins are usually highly cooperative processes, that is, they cannot occur in stepwise fashion. This implies that the binding of ligands to altered conformations (in an otherwise nativelike environment, where the protein molecule in the absence of binding would be in its native state) must also be a cooperative process. The binding of only one or a small number of ligands would not provide enough free energy to overcome the unfavorable free energy of the denaturation process if the latter can only occur altogether and not one step at a time. Cooperative binding of many ligand molecules can of course occur by other mechanisms as well, as will be discussed in Chapter 15, and the observation of cooperative binding should not be taken as necessarily indicative of binding to an altered conformational state. In general, optical and hydrodynamic measurements are required to demonstrate that a substantial conformational change has occurred.

It is appropriate to note in conclusion that lesser conformational changes exposing a small number of hydrophobic side chains may occur in some proteins. Dissociation of an oligomeric protein to its subunits, if it takes place without appreciable conformational change within the individual subunits, would be an example. Minor readjustments of surface groups are likely to occur whenever a complex ligand binds to a site on a protein molecule. These would not be detected by any but the most refined methods for examining protein structure, and will not be considered as representing significant conformational change as this term is used here.

REFERENCES

Cohn, E. J., and J. T. Edsall. (1943). *Proteins, Amino Acids and Peptides.* Reinhold, New York.

Jirgensons, B., and S. Capetillo. (1970). *Biochim. Biophys. Acta*, **214,** 1.

Nozaki, Y., and C. Tanford. (1971). *J. Biol. Chem.*, **246,** 2211.

Pauling, L., R. B. Corey, and H. R. Branson. (1951). *Proc. Nat. Acad. Sci. U.S.A.*, **37,** 205.

Tanford, C. (1962). *J. Am. Chem. Soc.*, **84,** 4260.

Tanford, C. (1968). *Adv. Protein Chem.*, **23,** 121.

Tanford, C. (1970). *Adv. Protein Chem.*, **24,** 1.

FIFTEEN

THE ASSOCIATION OF HYDROCARBONS AND AMPHIPHILES WITH COMMON SOLUBLE PROTEINS

As was pointed out in the preceding chapter, the existence of hydrophobic amino acid side chains on protein molecules suggests the possibility of interaction between proteins and small molecules containing hydrocarbon chains, such as hydrocarbons themselves, simple amphiphiles, and biological lipids. Interactions of this type presumably play a major role in the formation of functioning biological membranes but, before we proceed to discussion of membranes per se, it is worthwhile to examine available data on such interactions in simpler systems in which the protein is free in solution. This will be done in this and the two following chapters.

INTERPRETATION OF EXPERIMENTAL BINDING ISOTHERMS

The intrinsic information required for any investigation of the association between proteins and small molecules or ions consists of experimental data for the binding of the ligand to the protein, as a function of the equilibrium concentration of free ligand. We shall call the resulting curves, which of course are always obtained at constant temperature, pH, and other environmental factors, *binding isotherms*. A general discussion of the topic is provided by Steinhardt and Reynolds (1969).

As is well known, the thermodynamic interpretation of binding isotherms is never entirely unambiguous if more than one or two ligand molecules can be bound to a single protein molecule. If there are n discrete binding sites per protein molecule, each capable of binding ligand, and if these sites are all identical and independent, the average number of bound ligands per protein molecule ($\bar{\nu}$) is related to the equilibrium concentration of free ligand by the relation

$$\bar{\nu}/(n - \bar{\nu}) = KX_W f_W \tag{15-1}$$

where K is the equilibrium constant for binding at each site. All the data to be discussed refer to aqueous solutions and, for the reasons mentioned in Chapter 2, it is desirable to use mole fraction units for the free ligand concentration. Most original data are in molar units (C_W) and have been converted to mole fraction units to obtain the numerical data to be presented. In most cases the ligand concentrations are small and, in the absence of contrary evidence, f_W has been set equal to unity. The equilibrium constant in equation 15-1 is related to the standard free energy of transfer ($\mu_P^\circ - \mu_W^\circ$) of the ligand to a protein binding site by the relation

$$-RT \ln K = \mu_P^\circ - \mu_W^\circ \tag{15-2}$$

With free ligand concentration in mole fraction units, equation 15-2 gives the free energy difference in unitary units.

If equation 15-1 is not obeyed, and if the binding isotherm is *less steep* than it predicts, it means that a progressively smaller value of K must be used as X_W and $\bar{\nu}$ increase. This could mean that the n binding sites are nonidentical, possessing a range of intrinsic binding constants. On the other hand, a similar isotherm could result from n identical binding sites if there is interaction between sites with the result that ligand binding at any one site interferes with binding at other sites (e.g., by steric or electrostatic repulsion). In several of the examples to be cited below it is not possible to distinguish between these two interpretations (or some combination thereof) and we shall simply report an average value of $\mu_P^\circ - \mu_W^\circ$, corresponding to the value of K at the midpoint of the binding isotherm, that is, when $\bar{\nu} = n/2$.

If the binding isotherm is *steeper* than equation 15-1 predicts, cooperativity in the binding process is unequivocally indicated. There can be a variety of underlying causes. One possibility is the existence of discrete binding sites to which two or more ligand molecules must be bound simultaneously to achieve a significant gain in free energy, that is, the cooperativity may be an inherent property of the interaction of the ligand with the native protein. Another possibility is that the binding sites may be discrete and may only associate with one ligand each, but that a cooperative conformational change of the protein molecule may be necessary before the sites are available for binding, as discussed in the preceding chapter. Some of the data to be presented display extreme cooperativity in that all n sites are filled at virtually a single value of X_W. The binding phenomenon then becomes indistinguishable from micelle formation, that is, one possible interpretation is that the protein in this case is acting as a nucleus for deposition of ligand molecules in the form of a micellar shell. If this is the correct interpretation, the free energy change per ligand molecule entering the micellar shell is given by the analog of equation 7-1, that is,

$$\mu_P^\circ - \mu_W^\circ = RT \ln X_W + RT \ln f_W \tag{15-3}$$

It should be noted that equation 7-1 rather than equation 7-2 would be correct here because no new entropy of mixing term arises as a result of the micellar aggregation since the protein molecules are already dispersed in the solvent before association occurs. It should also be observed that the number of sites (n) does not appear in equation 15-3 and, therefore, need not be known to obtain $\mu_P^\circ - \mu_W^\circ$ from a highly cooperative isotherm.

Equation 15-3 is actually applicable *regardless of mechanism* if the isotherm is extremely steep. If ligand binding occurs noncooperatively to discrete sites, but a conformational change is required to make the sites available, the isotherm will become very steep if the free energy change $\Delta G_{\text{conf}}^\circ$ for the conformational change (in the absence of bound ligand) is sufficiently large (about 10 kcal/mole will suffice), and in that case equation 15-3 applies (Foster and Aoki, 1953, Reynolds et al., 1967), the only difference being that $\mu_P^\circ - \mu_W^\circ$ must now be considered as the sum of two terms,

$$\mu_P^\circ - \mu_W^\circ = (\mu_P^\circ - \mu_W^\circ)_{\text{int}} + \Delta G_{\text{conf}}^\circ / n \tag{15-4}$$

where $(\mu_P^\circ - \mu_W^\circ)_{\text{int}}$ is the average intrinsic free energy of association of the ligand with the n binding sites of the altered conformation. Since a conformational change may also be a prerequisite for micellelike association, an interpretation of $\mu_P^\circ - \mu_W^\circ$ for that mechanism may also require equation 15-4; $(\mu_P^\circ - \mu_W^\circ)_{\text{int}}$ in this case represents the free energy of incorporation of an amphiphile molecule into the micellar shell of the protein in its altered conformation.

Actual experimental binding isotherms often consist of successive stepwise increments in $\bar{\nu}$, as illustrated by Fig. 15-1. If all such steps were noncooperative and if there were no evidence for conformational change at any stage, the situation would be analogous to that existing in the binding of protons to proteins (Tanford, 1961, chapter 8), and the most likely interpretation would be that each step represents binding to a set of n_i discrete binding sites, each succeeding set having weaker affinity for the ligand. Apart from possible uncertainty in the evaluation of the precise values of n_1, n_2, and so on, because of overlap between the successive steps, the analysis

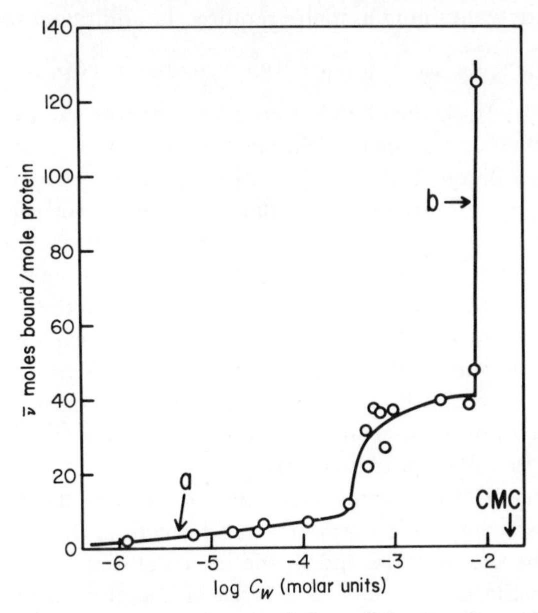

Fig. 15-1. Binding isotherm for the association of decyl sulfate with bovine serum albumin, at 2°C, pH 5.6, ionic strength 0.03. The number of amphiphile ions required to saturate the protein in the final cooperative process has not been established in this and most other isotherms discussed in the text. (It has been established for dodecyl sulfate under somewhat different conditions, as shown in Fig. 15-5.) Data used for this figure were obtained by equilibrium dialysis and are given in different form by Reynolds et al. (1967). Comparable studies of this type have been made with human serum albumin and indicate no significant differences from the bovine protein. The approximate critical micelle concentration of decyl sulfate under the experimental condition is shown at the lower right of the figure.

of the data from each step would be no more complicated than for an isotherm consisting of a single sigmoid curve. When some of the steps are cooperative and/or accompanied by conformational change that can be detected by optical and hydrodynamic methods, the simplest and customary procedure is still to treat each successive step as phenomenologically distinct. In this situation, however, one cannot know whether the increase in the number of binding sites in a particular step is to be considered a true increment (previously filled higher affinity sites remaining unaffected) or whether it represents a process in which sites of higher affinity are abolished and replaced by a larger number of more or less equivalent sites with weaker affinity for the ligand. In the latter case the interpretation of $\mathring{\mu}_P - \mathring{\mu}_W$ for a cooperative step involving the binding of n_2 moles of ligand to a protein molecule already containing n_1 moles requires an additional term, that is,

$$\mu_P^\circ - \mu_W^\circ = (\mu_P^\circ - \mu_W^\circ)_{int} + (n_1/n_2)\Delta(\mu_P^\circ - \mu_W^\circ)_{int} + (\Delta G_{conf}^\circ/n_2) \qquad (15\text{-}5)$$

where $\Delta(\mu_P^\circ - \mu_W^\circ)_{int}$ is the change in the average free energy of association of the previously bound n_1 ligands that accompanies the cooperative process.

Additional ambiguities in interpretation arise for cooperative steps of intermediate steepness. However, a value for $\mu_P^\circ - \mu_W^\circ$ for each step can always be obtained. Moreover, uncertainty in the interpretation of the numerical values of $\mu_P^\circ - \mu_W^\circ$ for a given amphiphile-protein system is not uniquely associated with complexity for, as was noted in the preceding chapter, diverse phenomena affect the value of $\mu_P^\circ - \mu_W^\circ$ even in the simplest situation, such as one would have if there is only a single binding site per protein molecule. Regardless of mechanism or complexity, it is generally not useful to attempt an interpretation unless one has available a large body of information for a variety of ligands.

Difficulties of this kind apply to a lesser degree to the free energies of simpler processes that were examined in Chapters 2, 3, and 7. What was found there was that a major aid in the interpretation of experimental results was the availability of data for a series of homologous molecules differing only in the length of the hydrocarbon chain. The same principle applies here. For amphiphiles, for example, interactions involving the hydrophilic head group may be assumed to be the same for a series of such ligands, so that the effect of hydrocarbon chain length may be taken to reflect solely the hydrophobic part of the interaction. If a conformational change accompanies the binding, there may be evidence that the structural change is the same for a series of ligands, in which case $\Delta G^\circ{}_{conf}$ (which refers to the free energy change in the absence of bound ligand) would be the same for all of them. In favorable circumstances the effect of hydrocarbon chain length on $\mu_P^\circ - \mu_W^\circ$ may thus primarily reflect the effect on $(\mu_P^\circ - \mu_W^\circ)_{int}$ and provide direct information on the hydrophobic component of the free energy.

COMPETITION BETWEEN BINDING AND SELF-ASSOCIATION

Competition between several ligands for the same protein binding sites is a familiar phenomenon that can be recognized without difficulty by varying the kind and amount of possible competing substances (e.g., buffer ions) in the medium. It can be avoided in the study of a particular ligand by eliminating competing substances from the medium. Substances that form complexes with the ligand, in competition with protein-ligand binding, can be dealt with in the same way. In the study of the binding of hydrophobic ligands there is an entirely different competing process: self-association, which cannot be avoided. Self-association may take the form of phase separation (for hydrocarbons or alcohols) or of micelle formation. In either case the value of X_W cannot be increased above a limiting value, and portions of the isotherm that lie at higher values of X_W cannot be experimentally observed. If the binding is noncooperative this means that only half or less of the isotherm can be observed unless $-(\overset{\circ}{\mu}_P - \overset{\circ}{\mu}_W)$ *exceeds* the free energy gain in the competing process (Fig. 15-2*a*); if the binding is highly cooperative it cannot be observed at all unless the free energy gain on binding to the protein exceeds that of the competing process (Fig. 15-2*b*).

It is evident that this competitive feature of the binding process effectively limits the acquisition of analyzable data to situations where the free energy gain associated with binding to the protein exceeds the free energy gain in competing processes. If a binding isotherm is observable at all, the affinity of the ligand for the protein must be high in comparison with other processes previously considered. If binding cannot be observed, we cannot say that there are no binding sites on the protein, but only that $\overset{\circ}{\mu}_P - \overset{\circ}{\mu}_W$ for binding sites is not negative enough to compete effectively with self-association processes. For cooperative processes less than 1 kcal/mole in $\overset{\circ}{\mu}_P - \overset{\circ}{\mu}_W$ can make the difference between ability to observe full saturation of the available sites and the absence of any observable binding at all.

BINDING OF HYDROCARBONS TO β-LACTOGLOBULIN AND OTHER PROTEINS

There have been only few measurements of the interaction between proteins and hydrocarbons. The most interesting study (Wishnia and Pinder, 1966)

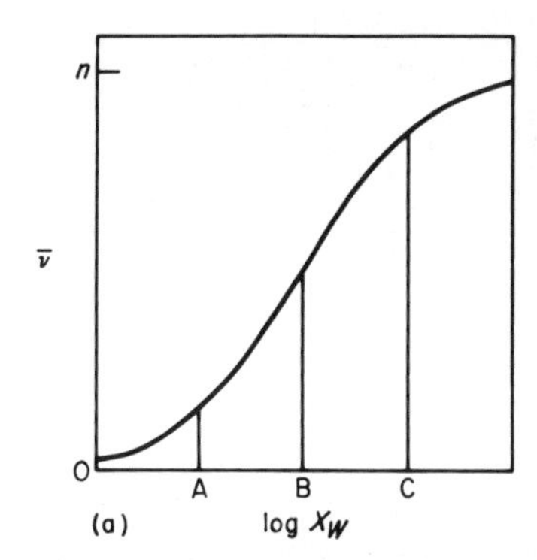

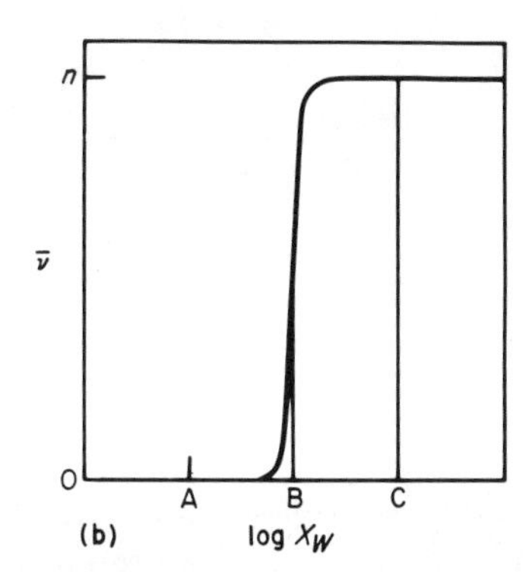

Fig. 15-2. Schematic diagrams of binding isotherms in relation to the maximum value of X_W attainable by virtue of self-association processes such as micelle formation. A represents the limiting value for X_W when $\overset{\circ}{\mu}_{mic} - \overset{\circ}{\mu}_W$ is about 1 kcal/mole more negative than $\overset{\circ}{\mu}_P - \overset{\circ}{\mu}_W$, B when the two are equal, and C when $\overset{\circ}{\mu}_{mic} - \overset{\circ}{\mu}_W$ is about 1 kcal/mole more positive than $\overset{\circ}{\mu}_P - \overset{\circ}{\mu}_W$. (*a*) Noncooperative; (*b*) highly cooperative.

involves the binding of butane and pentane to β-lactoglobulin, as measured from relative solubilities of these hydrocarbons in aqueous solutions with and without added protein. The protein was found capable of binding only two molecules of these hydrocarbons, and competitive binding studies showed that the same sites were used by both ligands. Equation 15-1 was obeyed for butane, suggesting that the two sites are identical, but the data for pentane required a lesser affinity for the second molecule than the first. The interpretation suggested by Wishnia and Pinder is that the protein has a single hydrophobic patch, able to accommodate hydrocarbon with a molar volume of 200 to 230 cc. This provides a comfortable fit for two butane molecules, but a rather tight one for two pentane molecules. Nonadherence of the binding of the latter to equation 15-1 is thus ascribed to steric interference with binding of the second pentane molecule, the two halves of the hydrophobic patch constituting intrinsically identical sites. With this assumption, $\mu_P^\circ - \mu_W^\circ$ values can be obtained for pentane as well as butane by equation 15-1, provided one limits oneself in the case of pentane to results at low $\bar{\nu}$ where the fraction of protein molecules with two bound ligands is insignificant. The results obtained are shown in Table 15-1. The interpretation in terms of a single hydrophobic binding site is supported by the fact that β-lactoglobulin binds one molecule of dodecyl sulfate with very high affinity (Ray, 1968), roughly consistent with the value of $\mu_P^\circ - \mu_W^\circ$ expected on the basis of Table 15-1 for a molecule of *n*-decane.

Table 15-1. Free Energy of Association of Hydrocarbons with β-Lactoglobulin at 25°C[a]

	$\mu_P^\circ - \mu_W^\circ$ (kcal/mole)	$\mu_{HC}^\circ - \mu_W^\circ$[b] (kcal/mole)
Butane	−6.44	−5.97
Pentane	−7.14	−6.85

[a] From Wishnia and Pinder (1966).
[b] From equation 2-3.

In Table 15-1 the values of $\mu_P^\circ - \mu_W^\circ$ are compared with the values of $\mu_{HC}^\circ - \mu_W^\circ$ for the same hydrocarbons, taken from Table 4-1. It is seen that $\mu_P^\circ - \mu_W^\circ$ is somewhat more negative, as it has to be, since experimental data in excess of 50% saturation were obtained. In molecular terms, this presumably means that the process of binding not only removes the bound hydrocarbon from an aqueous environment, but also has a beneficial effect on the apolar side chains of the protein. One model that would satisfy this requirement is a cleft in the protein surface which, in the absence of bound hydrocarbon, possesses a hydrocarbon-water interface. One would expect that the protein-bound hydrocarbon might be more rigid than the dissolved molecule and this alone would make $\mu_P^\circ - \mu_W^\circ$ more positive than $\mu_{HC}^\circ - \mu_W^\circ$: the fact that the difference between $\mu_P^\circ - \mu_W^\circ$ for butane and pentane is somewhat less than the difference in $\mu_{HC}^\circ - \mu_W^\circ$ is perhaps an indication of this. This and other minor factors mentioned on page 123 all lead to the conclusion that the difference of 0.3 to 0.5 kcal/mole between μ_P° and μ_{HC}° in Table 15-1 gives only a lower limit for the contribution of the amino acid side chains of the protein to $\mu_P^\circ - \mu_W^\circ$: the actual contribution is likely to be higher, say $\sim$1 kcal/mole. Even this would be a small figure, less than would be contributed by a single leucine or valine residue completely surrounded by water (Table 14-1). Thus only partial exposure of hydrophobic side chains at the binding site is required to account for the experimental data, and the most noteworthy feature of the binding site is the limitation on the volume of hydrocarbon it can accommodate.

Wishnia and Pinder (1966) have also determined $\bar{H}_P^\circ - \bar{H}_W^\circ$ and $\bar{S}_P^\circ - \bar{S}_W^\circ$ for this system by measuring binding at two temperatures. The values obtained for butane, -1.1 kcal/mole and $+18$ cal/mole-deg, respectively, are of the same order of magnitude as similar data for the transfer of butane from water to pure hydrocarbon (Table 4-1) or to the interior of a micelle (Table 6-1). Closely similar values were obtained for pentane. It is doubtful whether the differences among these data are experimentally significant: $\bar{S}_P^\circ - \bar{S}_W^\circ$ is not as positive as might have been anticipated, especially if we take into account the expected contribution from the hydrophobic residues of the protein binding site, and this deviation is consistent with the probability that binding to the protein is accompanied by some decrease in flexibility.

Among other proteins that have been investigated, ribonuclease and lysozyme (Wishnia, 1962) do not bind hydrocarbon at all at the hydrocarbon concentration to which one is limited. Serum albumin can bind several molecules of small hydrocarbons (Wetlaufer and Lovrien, 1964; Wishnia and Pinder, 1966), but the isotherms have the form illustrated by Fig. 15-2*a*, with the cutoff at point A, indicating that the binding reflects low levels of saturation of a fairly large number of sites of weak affinity for hydrocarbon. It is clear that hydrophobic surface patches of even such limited size as that possessed by β-lactoglobulin do not occur commonly on water-soluble proteins.

BINDING OF AMPHIPHILES TO SERUM ALBUMIN AND OTHER PROTEINS

There is a large body of experimental data on the binding of amphiphiles to proteins, which has been reviewed in detail by Steinhardt and Reynolds

(1969). The majority of such studies have involved the binding of anionic amphiphiles to a single protein, serum albumin: this protein functions biologically as a carrier for fatty acid anions and other simple amphiphiles. Experimental studies of the association, as illustrated by Fig. 15-1, have generally yielded complex results, indicating several levels of affinity. The limited data available for other proteins indicate that binding with very high affinity may be nearly unique to serum albumin: several other common water-soluble proteins do not show it. The lowest affinity type of association, however, which proves to be a cooperative process in which protein can bind more than its own weight of amphiphile, appears to be a nonspecific process common to all proteins. We shall confine the discussion here to these extremes of the binding isotherm (i.e., regions *a* and *b* of Fig. 15-1). It will be seen that the product of the final cooperative stage (region *b*) may represent two quite different types of complexes, depending on the hydrocarbon chain length of the amphiphilic anion.

High Affinity Sites of Native Serum Albumin

There are about 10 or 11 discrete binding sites with a high association constant for anionic amphiphiles, and considerable information concerning them is available, most of it from the work of Reynolds et al. (1967, 1968, 1970). The sites are quite specific for ligands with anionic head groups. Hydrocarbons alone do not bind strongly to serum albumin, as previously mentioned. Cationic amphiphiles are unable to bind to most of the binding sites.[1] Aliphatic alcohols bind to about half the sites, but with considerably less affinity than anionic molecules with hydrocarbon chains of the same length (Reynolds et al., 1968). The best evidence for active involvement of the anionic head group and selectivity for specific head groups is provided by the finding of Reynolds et al. (1968) that some of the binding sites, identifiable by perturbation of the tryptophan spectrum when binding takes place, do not bind alkyl carboxylates at all (or do so with greatly diminished affinity), whereas both alkyl sulfates and sulfonates are strongly bound. As previously noted, no such head group specificity is observed in micelle formation.

It is well known that serum albumin is capable of combining with a large variety of anions, including such simple ions as Cl^-, SCN^-, and acetate. Quantitative binding isotherms were published as early as 1950 (Scatchard

[1] Work in progress in the author's laboratory (Y. Nozaki, unpublished data) indicates that there may be four binding sites with moderately high affinity for cationic amphiphiles. The hydrophobic areas of these sites may be shared with some of the sites for anionic amphiphiles, as illustrated schematically in Fig. 15-4.

et al., 1950), and showed that a large number of sites were available on the protein, but that about 10 of these have higher affinity than the rest. No proof exists that these are the same sites as those to which anionic amphiphiles bind, but the numerical coincidence may not be fortuitous. It should be noted that the binding constants for these simple ions are several orders of magnitude smaller than those for binding of amphiphilic anions and that the separation of sites into affinity classes is more difficult. Later data of higher precision led to the conclusion that there are only nine high affinity sites (Scatchard et al., 1957, 1959) and the most recent data (Scatchard and Yap, 1964) reduce the number to five. The latter data were confined to measurements at low concentrations and do not preclude the possible existence of another set of four or five sites of weaker, but still relatively strong affinity for anions. All of the later measurements indicate that one of the sites we have referred to as high affinity sites is a very much stronger binding site than the rest, and that binding to this site is remarkably endothermic. There is a similar unique site for the binding of anionic amphiphiles (Goodman, 1958), and the ΔH for association of dodecyl sulfate with this site is -18 kcal/mole (Lovrien and Sturtevant, 1971).

In calculating $\mu_{P}^{\circ} - \mu_{W}^{\circ}$ for the binding of amphiphiles we have ignored the one exceptionally strong site because accurate data for this site (i.e., precise X_W values at $\bar{\nu}$ values <1) are extremely difficult to measure and have not been obtained. Average values of $\mu_{P}^{\circ} - \mu_{W}^{\circ}$ (page 127) have been calculated for the remaining sites, and some of them are shown in Fig. 15-3. They are of course much more negative than the values of $\mu_{mic}^{\circ} - \mu_{W}^{\circ}$ for the same substances, as is evident from a comparison (Fig. 15-1) of the free ligand concentration required to saturate the high energy sites and the critical micelle concentration. This greater free energy gain is ascribable to the contribution from the interaction between the amphiphile head group and the protein's anion-binding sites, a conclusion required by the experimental data regardless of whether the binding sites here have a one-to-one correspondence with the *strongest* general anion binding sites. It is not unreasonable to assign 3 to 4 kcal/mole to this interaction, which would mean that only about one half of the value of $\mu_{P}^{\circ} - \mu_{W}^{\circ}$ may result from interaction with hydrophobic surface patches of the protein. This, and the inability to bind hydrocarbons such as butane and pentane to a significant degree, may indicate that the hydrophobic part of the binding site may not be a single contiguous area, but that it may involve separated areas such as illustrated schematically by Fig. 15-4. Binding to a site of this type would involve an "inch-worm" type of attachment of the ligand. Other evidence supporting a binding site such as is illustrated by Fig. 15-4 is provided by the fact that oleate (containing one centrally located double bond) binds more strongly to some of the binding sites than its fully saturated analog, stearate (Good-

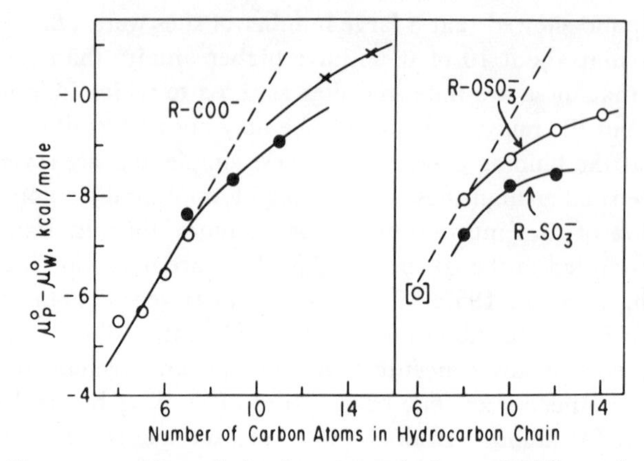

Fig. 15-3. Free energy of association between anionic amphiphiles and native bovine serum albumin (from Tanford, 1972). For $RCOO^-$ ○ data of Teresi and Luck (1952), 1°C, ionic strength 0.2; ● data of Reynolds et al., (1968), 2°C ionic strength 0.03; X data of Spector et al., (1971), 37°C, ionic strength 0.15. For $ROSO_3^-$ and RSO_3^- all data are from Reynolds et al., (1967), 2°C, ionic strength 0.03, except that the point for hexyl sulfate is a result of poorer precision, taken from Steinhardt and Reynolds (1969).The dashed lines show the *slopes* of plots of $\mu_{HC}^{\circ} - \mu_W^{\circ}$ or $\mu_{ROH}^{\circ} - \mu_W^{\circ}$ versus n_C as given in Chapter 3.

man, 1958; Spector et al., 1971), even though double bonds decrease hydrophobicity in general (Chapter 2), and $\mu_{HC}^{\circ} - \mu_W^{\circ}$ is less negative for oleate than for stearate (page 14). This may indicate that the central portion of the hydrocarbon chain does not contribute to the free energy of binding and that the hydrophobic binding areas are arranged so that a bend in the chain at the center, such as a double bond would introduce without loss of rotational flexibility, is helpful in the binding process.

Regardless of the exact nature of the binding process, the effect of chain length on $\mu_P^{\circ} - \mu_W^{\circ}$, as shown by Fig. 15-3, unequivocally demonstrates that the hydrophobic parts of the binding sites are circumscribed and limited in the length of hydrocarbon chain that can be accommodated. The dashed lines of Fig. 15-3 are drawn with a slope equal to the slopes of Fig. 3-1 and represent the expected effect of hydrocarbon chain length on $\mu_P^{\circ} - \mu_W^{\circ}$ arising from removal of the hydrocarbon chain from water. The experimental dependence of $\mu_P^{\circ} - \mu_W^{\circ}$ on n_C is seen to be of the right order of magnitude up to $n_C \simeq 8$, but a pronounced decrease is observed for $n_C > 8$. It evidently becomes increasingly difficult to find suitable locations for the hydrocarbon chain as its length increases, and little or no gain in free energy of binding occurs above $n_C = 12$.

It is of interest, in conclusion, to note the displacement by about one

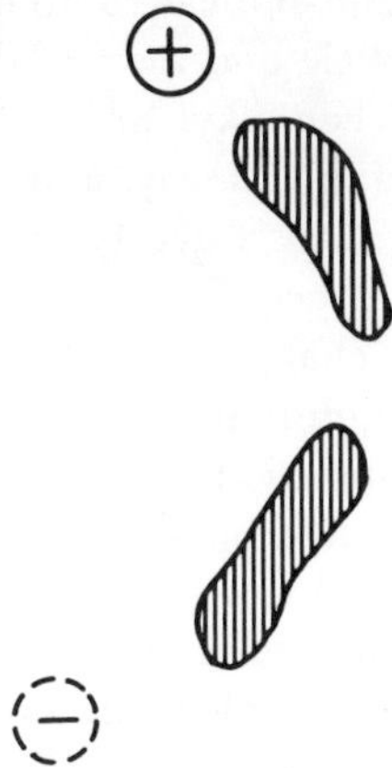

Fig. 15-4. Schematic diagram of a binding site of native serum albumin for amphiphiles. The hydrophobic patch (hatched area) is shown as broken into regions because neither butane nor pentane can be bound effectively. The two regions are placed at an angle to each other because at least one unsaturated fatty acid anion is bound with greater affinity than its saturated analog. The site marked with a plus sign represents one of the general anion binding sites of serum albumin: its juxtaposition to the hydrophobic patches is an essential element of the site to account for the fact that anionic amphiphiles are bound strongly whereas hydrocarbons are not. A small number of binding sites with moderate affinity for cationic amphiphiles exists, indicating that some of the binding sites may also have a negatively charged position near the hydrophobic areas as shown by the circle marked with a minus sign.

carbon atom of the positions of the curves for alkyl sulfates and sulfonates in Fig. 15-3. The underlying reason is presumably the same as that for the similar difference in the cmc values for these molecules, as discussed on page 52.

Massive Cooperative Binding Without Significant Conformational Change

When binding measurements are extended to amphiphile concentrations above those required to saturate the discrete sites of serum albumin, further association occurs. The final step is always a massive cooperative process in which a large number of amphiphile molecules are incorporated over a very narrow range of free amphiphile concentration, as shown by region *b* of Fig. 15-1. In the association of octyl and decyl sulfates and of octyl, decyl, and dodecyl sulfonates with bovine serum albumin (Reynolds et al., 1967) the cooperative process occurs without drastic conformational change, as judged by intrinsic viscosity and optical rotation, that is, the product is a compact, nearly spherical particle and most of the internal structure of the native molecule is retained. When alkyl sulfates and sulfonates with longer hydrocarbon chains are bound in this way, a dramatic change in conformation accompanies the process, as discussed below.

A remarkable aspect of the process, for those amphiphiles for which it is not accompanied by a major change in conformation, is that it occurs at essentially the same critical amphiphile concentration for all of them. If we calculate the free energy of association by equation 15-3, $\mu_P^\circ - \mu_W^\circ$ is clearly independent of the length of the hydrocarbon chain (from $n_C = 8$ to $n_C = 12$ in the case of the sulfonates). Whatever the mechanism of association may be,[2] an increase in hydrocarbon chain length beyond $n_C = 8$ cannot contribute to the stability of the complex, although it does not prevent its formation.

There is evidence that all proteins are capable of forming complexes of this type with the same anionic amphiphiles as were used for serum albumin (J. A. Reynolds, personal communication). No similar studies with cationic amphiphiles are available. The problem deserves further study.

Massive Cooperative Binding Accompanied by Drastic Conformational Change

The cooperative formation of stoichiometric complexes between serum albumin and dodecyl sulfate, containing a very large number of bound ligands, was first recognized by Putnam and Neurath (1945). They and similar complexes formed by myristyl (tetradecyl) sulfate and other detergents with long hydrocarbon chains are quite different from the complexes formed by the shorter alkyl sulfates or the alkyl sulfonates (to $n_C = 12$) because large changes in optical and hydrodynamic properties, indicative of a major conformational change, accompany their formation (Reynolds et al., 1967; Steinhardt et al., 1971). Another important difference is that there is no limitation in the length of hydrocarbon chain (at least to $n_C = 14$) that can be accommodated in this type of complex with a gain in free energy.

[2] Since there is no independent evidence for a significant conformational change, formation of a micellar shell about the native protein is the most likely mechanism. In terms of equation 15-5, $n_2 \gg n_1$, so that the second term on the right-hand side of the equation makes only a small contribution to $\mu_P^\circ - \mu_W^\circ$. The last term would be zero if the proposed micellelike binding mechanism is correct, and would in any event not be expected to contribute much to the effect of hydrocarbon chain length. It is difficult to imagine that the lack of dependence of $\mu_P^\circ - \mu_W^\circ$ on n_C° implies anything other than the lack of a significant effect of hydrocarbon chain length on the intrinsic free energy of association between amphiphile and protein.

The values of $\mu_P^\circ - \mu_W^\circ$, determined by equation 15-3, at 2°C, pH 5.6, ionic strength 0.03, are -8.0 ± 0.5 kcal/mole for myristyl sulfate and -6.8 ± 0.5 kcal/mole for dodecyl sulfate. A minimal value for decyl sulfate can be obtained since decyl sulfate forms the compact type of cooperative complex in preference to the type being considered here. This requires that $\mu_P^\circ - \mu_W^\circ$ for the latter must be less negative than $\mu_P^\circ - \mu_W^\circ$ for the compact complex which, under the same conditions, is -5.3 ± 0.5 kcal/mole. It follows that the increment in $\mu_P^\circ - \mu_W^\circ$ per carbon atom is in the range of -500 to -1000 kcal/mole, as expected for the effect of hydrocarbon chain length on removal of the amphiphile from water to a hydrophobic site.

It has been established (for dodecyl sulfate) that this type of complex can be formed with all proteins, and that the maximum amount of ligand that can be bound per gram of protein is the same for all of them (Pitt–Rivers and Impiombato, 1968; Reynolds and Tanford, 1970a). For proteins without disulfide bonds, or with disulfide bonds reduced if originally present, there are actually two forms of the complex, as can be seen from Fig. 15-5, one containing 0.4 g dodecyl sulfate per g protein (one amphiphile per seven amino acid residues), the other 1.4 g dodecyl sulfate per g protein (one amphiphile per two amino acid residues). The transition between them takes place at 6 to $7 \times 10^{-4} M$ dodecyl sulfate at 25°C, pH 5.6 to 7.2, with little dependence on ionic strength. All proteins are dissociated to their constituent polypeptide chains, and each chain forms an individual asymmetric particle, approximately rodshaped, with length roughly proportional to molecular weight (Reynolds and Tanford, 1970b). A model in which the dodecyl sulfate forms a cylindrical micellar shell about a somewhat flexible extended protein core would be consistent with the results[3] and, if correct, would suggest that the two forms of the complex might represent different ways of packing the detergent about the protein core. Results obtained for proteins with intact disulfide bonds indicate that very similar complexes are formed, subject only to steric limitations imposed by the disulfide bonds. The amount of detergent bound is diminished (Pitt–Rivers and Impiombato, 1968) and the resulting complex is not as extended (Fish et al., 1970). There is evidence that a similar complex can be formed with cationic amphiphiles, as is reasonable on the basis of this model, in which the head groups are not directly involved, but detailed data are not yet available.

[3] We should emphasize, as pointed out in the early part of this chapter, that no distinction between micellelike binding and single site binding to the altered protein can be made when clear evidence for a conformational change exists. The surface area of the rodlike complexes is larger than one would expect for an uninterrupted micellar shell, given the binding ratio of 1.4 g dodecyl sulfate per g of protein, if the surface area per head group is to be similar to that of pure micelles as given in Chapter 9. The discrepancy would be even greater at the lower binding level of 0.4 g per g of protein.

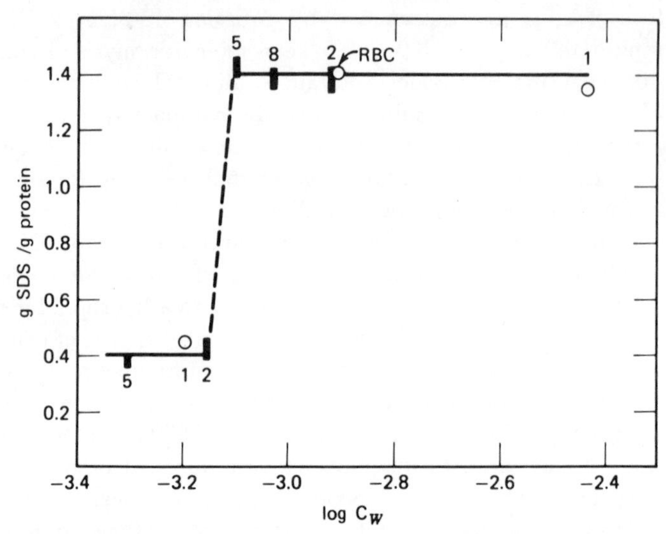

Fig. 15-5. Binding of dodecyl sulfate to proteins at 20°C, with disulfide bonds reduced where present in the native state. Similar results were obtained for a large variety of proteins, and the number adjacent to each point represents the number of different proteins used to obtain the result at each value of C_W. The point labeled "RBC" represents a mixture of proteins extracted from erythrocyte membranes. Measurements were made over a broad range of ionic strength with consequent variation in the cmc of the detergent. Open circles represent results where C_W is below the cmc. Filled rectangles represent results where the total concentration of uncombined dodecyl sulfate is above the cmc, so that micelles were present: C_W represents the concentration of monomeric ligand, equal to the cmc at the ionic strength employed. The half-filled rectangle represents experimental data with and without micelles at the same value of C_W (from Reynolds and Tanford, 1970a).

ASSOCIATION BETWEEN PROTEINS AND AMPHIPHILE MICELLES

In the foregoing discussion we have considered only the binding of amphiphile to single protein molecules, at free amphiphile concentrations below the cmc. An entirely different type of association is possible: the binding of protein molecules to preexisting amphiphile micelles. It has been conclusively demonstrated (Reynolds and Tanford, 1970a; see Fig. 15-5) that such association does not occur between proteins and dodecyl sulfate micelles when both components are present at relatively low concentrations. It is important to recognize that this result cannot be used as evidence for a general extension to all micellar systems because the interaction between proteins and dodecyl sulfate precludes the coexistence of *native* proteins and micelles, that is, as we have shown, all proteins form micellelike complexes with dodecyl sulfate at amphiphile concentrations below the cmc. At sufficiently high dodecyl sulfate concentrations for the simultaneous existence of dodecyl sulfate micelles, the protein itself is therefore already in the form of a complex with high dodecyl sulfate content and high charge density of the same sign as that of the micelles. The conditions are not favorable for further interaction unless the total concentration of both solutes is increased by orders of magnitude to the point where dodecyl sulfate micelles would themselves tend to interact to form liquid crystalline structures (page 73).

CONCLUSIONS

The most interesting aspects of the results discussed in this chapter are (1) the element of competition between self-association and binding to protein, (2) the limitation on the length of hydrocarbon chain that can be utilized for at least some of the modes of association with protein, and (3) the importance of head group interactions in the binding of amphiphiles to native serum albumin.

Combination of the first two of these phenomena indicates that competition between the binding of amphiphilic ligands to native serum albumin and their incorporation into micelles will shift in favor of micelle formation as the size of the hydrophobic portion of the ligand molecule increases. This is likely to be an important factor in the biological function of serum albumin as a carrier protein, because it prevents the protein from combining with anionic phospholipids (except at the one very strong binding site for which exact thermodynamic data are not available). The affinity for phospholipids, in terms of a binding site such as illustrated by Fig. 15-3, will be no greater than the affinity for a fatty acid anion, whereas $\mu^{\circ}_{\text{mic}} - \mu^{\circ}_{W}$ is much more negative for the phospholipid (Table 12-2).

The interplay of size limitation and competition also enters into the possible association between native proteins and amphiphile micelles. If there is a size limit in the ability of amphiphiles to form rodlike micellar complexes with proteins, such complexes will not form below the cmc for phospholipids and other amphiphiles with very large hydrophobic moieties. This would lead to the coexistence of native proteins with micelles and allow for association between the protein and the micelle. It will be seen in the following chapter that such association can be observed between proteins and micellar aggregates of phospholipids with two hydrocarbon chains. This indicates that the rodshaped cooperative complex between proteins and dodecyl sulfate, for which no size limitation could be detected on the basis of the limited data presented here, is in fact subject to at least some restriction.

The third factor considered above, the involvement of head groups in the binding of amphiphiles to proteins is perhaps too obvious to need emphasis. It allows for specificity in the association of proteins with amphiphiles even though the major driving force for the association may be the hydrophobic interaction, which itself is nonspecific.

REFERENCES

Fish, W. W., J. A. Reynolds, and C. Tanford. (1970). *J. Biol. Chem.*, **245,** 5166.
Foster, J. F., and K. Aoki. (1958). *J. Am. Chem. Soc.*, **80,** 5215.
Goodman, D. S. (1958). *J. Am. Chem. Soc.*, **80,** 3892.
Lovrien, R., and J. M. Sturtevant. (1971). *Biochemistry*, **10,** 3811.
Pitt-Rivers, R., and F. S. A. Impiombato. (1968). *Biochem. J.*, **109,** 825.
Putnam, F. W., and H. Neurath. (1945). *J. Biol. Chem.*, **159,** 195.
Ray, A. (1968). In *Solution Properties of Natural Polymers*, Special Publication No. 23, The Chemical Society, London.
Reynolds, J. A., and C. Tanford. (1970a). *Proc. Nat. Acad. Sci. U.S.A.*, **66,** 1002.
Reynolds, J. A., and C. Tanford. (1970b). *J. Biol. Chem.*, **245,** 5161.
Reynolds, J. A., S. Herbert, H. Polet, and J. Steinhardt. (1967). *Biochemistry*, **6,** 937.
Reynolds, J., S. Herbert, and J. Steinhardt. (1968). *Biochemistry*, **7,** 1357.
Reynolds, J. A., J. P. Gallagher, and J. Steinhardt. (1970). *Biochemistry*, **9,** 1232.
Scatchard, G., and W. T. Yap. (1964). *J. Am. Chem. Soc.*, **86,** 3434.
Scatchard, G., J. S. Coleman, and A. L. Shen. (1957). *J. Am. Chem. Soc.*, **79,** 12.
Scatchard, G., Y. V. Wu, and A. L. Shen. (1959). *J. Am. Chem. Soc.*, **81,** 6104.
Scatchard, G., I. H. Scheinberg, and S. H. Armstrong. (1950). *J. Am. Chem. Soc.*, **72,** 535, 540.
Spector, A. A., J. E. Fletcher, and J. D. Ashbrook. (1971). *Biochemistry*, **10,** 3229.
Steinhardt, J., J. Krijn, and J. G. Leidy. (1971). *Biochemistry*, **10,** 4005.
Steinhardt, J., and J. A. Reynolds. (1969). *Multiple Equilibria in Proteins*, Academic Press, New York.
Tanford, C. (1961). *Physical Chemistry of Macromolecules.* John Wiley and Sons, New York.
Tanford, C. (1972). *J. Mol. Biol.* **67,** 59.
Teresi, J. D., and J. M. Luck. (1952). *J. Biol. Chem.*, **194,** 823.
Wetlaufer, D. B., and R. Lovrien. (1964). *J. Biol. Chem.*, **239,** 596.
Wishnia, A. (1962). *Proc. Nat. Acad. Sci. U.S.A.*, **48,** 2200.
Wishnia, A., and T. W. Pinder, Jr. (1964). *Biochemistry*, **3,** 1377.
Wishnia, A., and T. W. Pinder, Jr. (1966). *Biochemistry*, **5,** 1534.

SIXTEEN

ASSOCIATION OF PROTEINS WITH BIOLOGICAL LIPIDS

A cell membrane is the complex product resulting from the association of specific proteins with biological lipids, but the mechanism by which this association occurs is not yet understood. The properties of specific membrane proteins and the procedures that have to be used to dissociate them from membranes suggest that several different modes of association are possible, for example, some proteins are attached to the surface of the membrane and others may penetrate from one side to the other. (Information of this kind will be considered in chapter 19.) Ultimately such information will have to be interpreted in terms of specific protein–lipid interactions, that is, in terms of quantitative studies of the kind presented in the previous chapter for the association between proteins and simple amphiphiles.

Unfortunately, very little quantitative work on this subject is available, chiefly because of the experimental difficulties that are involved. The major underlying cause is the competitive aspect of protein–amphiphile interaction, discussed in the previous chapter. Thus the simplest type of protein–lipid interaction, the binding of single lipid molecules to specific binding sites of native protein molecules, is experimentally observable only if the free energy gained by lipid–protein association exceeds the free energy of self-association to form micellar aggregates. This creates two problems:

1. Because the cmc of many lipids of interest is expected to be about 10^{-9} to $10^{-10}M$ (Table 12-2), it is evident from Fig. 15-2 that binding isotherms would have to be conducted at extremely low free lipid concentrations. These concentrations can be measured only if radioactively labeled com-

pounds are used, and the acquisition of experimental data is thus limited by the availability of pure lipids containing appropriate isotopic labels.

2. We cannot expect common water-soluble proteins to have binding sites of the requisite affinity for association with lipids in this manner. Even serum albumin, with its exceptional affinity for amphiphile ions, has binding sites that are very limited in the size of the hydrocarbon moiety they can accommodate, as illustrated schematically by Fig. 15-4. Unless the hydrophobic parts of two sites of this kind happen to be close to each other there would be no way to locate a second hydrocarbon chain near the point of attachment of the head group, and even anionic phospholipids with two hydrocarbon chains would then bind to the protein with an affinity no greater than that of, say, dodecyl sulfate, and that would not suffice to compete with self-association of the lipid. One therefore is forced to use membrane proteins per se, or apoproteins from soluble lipoproteins, for an investigation of specific sites for association with lipid, and one is limited not only by the present lack of methods for preparing such proteins in pure form, but also by the fact that membrane proteins frequently have poor solubility in aqueous media.

To demonstrate the existence of highly cooperative association between a protein molecule and a large number of lipid molecules would be an easier problem. Complexes of this kind have to be formed at lipid concentrations below the cmc to be observable but, if formed, should be easily detectable because of their high lipid-to-protein ratio. The ability to observe association between native proteins and phospholipid *micelles* (see below) indicates that such complexes are not as readily formed by phospholipids with two hydrocarbon chains as they are by amphiphiles with a single hydrocarbon chain, such as the alkyl sulfates and alkyl sulfonates. Since this conclusion is based on experiments with proteins that are not typical membrane proteins, it may not be universally applicable. As Fig. 15-2*b* shows, only a small change in the relation between $\mu_P^\circ - \mu_W^\circ$ and $\mu_{mic}^\circ - \mu_W^\circ$ would be needed to differentiate proteins able and unable to form highly cooperative complexes, and such a difference could arise, for example, from a difference in the free energy of the conformational change required to permit formation of the complex. There are, however, other considerations, to be discussed in Chapter 19, that indicate that the suggestion made earlier (Reynolds and Tanford, 1970), that the cooperative complexes between proteins and dodecyl sulfate can serve as a model for a possible ubiquitous mode of association between proteins and lipids, is probably invalid.

One example of what appears to be nonspecific association between protein and lipids, indistinguishable from association with simpler amphiphiles, and mediated solely by hydrophobic forces, is provided by the work of Ji and Benson (1968). They measured the binding of palmitic acid and of three lipids to a protein (or, more likely, a mixture of proteins) from spinach chloroplasts. The lipids were among those naturally present in spinach leaves: two were glycolipids with neutral head groups and the other was a phosphatidylglycerol with a negatively charged head group, and all have two hydrocarbon chains per molecule. Binding isotherms (in 20% methanol) indicated that all three lipids, as well as palmitic acid, have similar affinity for the protein, and that saturation is reached in each case when about 30 hydrocarbon chains are bound to 23,000 grams of protein. This is a similar level of binding to that observed for one of the dodecyl sulfate–protein complexes in Fig. 15-5 (0.4 g dodecyl sulfate per g of protein, corresponding to one hydrocarbon chain per 7 amino acid residues), but it is unlikely that a similar type of complex is involved. The binding isotherms are only slightly cooperative, and indicate that a relatively large number of discrete and weakly interacting binding sites may be involved. The same protein has a single specific binding site for β-carotene, which is a pure hydrocarbon with a system of conjugated double bonds. (It should be noted that there is some question about whether true equilibrium was measured in these experiments, so that the results cannot be accepted without reservations.)

As was pointed out in the previous chapter, the absence of micellelike complexes between a protein molecule and a large number of amphiphilic ligands makes it possible for native proteins and amphiphile micelles to coexist, and this allows for a kind of association that does not occur with simple amphiphiles; the association of protein molecules with preformed micelles. Such association has been observed for several proteins, but in all cases reported thus far, it seems to involve primarily a nonspecific electrostatic interaction between proteins with a high positive charge and the negatively charged surface of a bilayer or monolayer formed by anionic phospholipids.

The most extensive studies have been carried out with cytochrome *c*, which bears about 10 positive charges per molecule at neutral pH. It has been shown to bind to monolayers, vesicles, and artificial membranes (e.g., Das et al., 1965; Kimelberg et al., 1970; Steinemann and Läuger, 1971), provided that these are composed of or contain anionic lipids. Cytochrome *c* can be incorporated in the internal aqueous compartments of vesicles formed by zwitterionic lipids (Papahdjopoulos and Watkins, 1967), but there is no evidence suggesting appreciable binding. The association with anionic lipids appears able to accommodate any number of protein molecules, an upper limit being reached only when the net charge of the bound protein neutralizes that of the lipid aggregate. The association is progressively disrupted by ionic strength, and especially by multivalent cations. X-ray diffraction studies of liquid crystalline structures containing cytochrome *c* (Shipley et al., 1969a; Gulik–Krzywicki et al., 1969) show that the protein forms a layer in the aqueous region between bilayers. All these results point to nonspecific association between the protein and the bilayer surface, and this is supported by the fact that other basic proteins, such as lysozyme and ribonuclease can form similar complexes (e.g., Gulik–Krzywicki et al., 1969; Kimelberg and Papahdjopoulos, 1971) and by the formation of a similar complex between cytochrome *c* and the phosphoprotein phosvitin, which has a high negative surface charge density by virtue of a high content of phosphoamino acids, but contains no lipid (Taborsky, 1970). Evidence suggesting possible secondary effects due to nonelectrostatic interactions is given in some of these studies, but the evidence is indirect and not conclusive.

Serum albumin has been shown to associate with multilayered vesicles formed from a mixture of phosphatidylcholine, cholesterol, and dicetyl phosphate (Sweet and Zull, 1970). Appreciable binding is observed only below the isoelectric pH of the protein, that is, where it bears a positive charge, and it increases in proportion to the negative charge on the vesicle, as provided by increasing amounts of the dicetyl phosphate.

An interesting feature of the cytochrome *c* complexes is that they can be extracted by organic solvents, isooctane having been commonly used for this purpose (Das and Crane, 1964). The extracted complexes are neutral and exist in the organic solvent in aggregated form, with perhaps 30 to 40 protein molecules per complex (Shipley et al., 1969b). There is no way in which such a hydrocarbon-soluble complex can be formed without disruption of the lipid bilayer structure, at least at the surface of the complex, so that hydrocarbon chains are at the interface with the solvent. The formation of such an inverted structure in this simple system has to be kept in mind when possible modes of interaction between proteins and lipids in biological membranes are considered.

Special mention should perhaps be made of the basic protein from myelin sheath (see Chapter 18), which forms similar complexes with anionic lipids and again not with zwitterionic lipids (Palmer and Dawson, 1969). The complexes are soluble in a mixture of chloroform and methanol, a mixture in which most of the myelin membrane is also soluble. This may suggest that the manner of attachment of this protein to the membrane in the myelin sheath is largely the result of simple electrostatic interaction.

REACTIVATION OF MEMBRANE-BOUND ENZYMES

Qualitative information about the interaction with lipids can be obtained for membrane *enzymes* on the basis of lipid requirements for expression of catalytic activity. One of the earliest and best examples is provided by the mitochondrial enzyme β-hydroxybutyrate dehydrogenase (Sekuzu et al., 1963; Fleischer et al., 1966). This enzyme can be solubilized and partially purified by use of the bile acid anion cholate, which acts as a mild detergent, or by treatment of mitochondria with phospholipase A. The latter converts phospholipids into lysophosphatidyl glycerides, which also have detergent-like properties. The solubilized enzyme has no activity, but incubation with suspensions of phosphatidylcholine from natural sources (presumably containing vesicles bounded by single and/or multiple lipid bilayers) restores activity. Phosphatidylethanolamine, cardiolipin, and other mitochondrial lipids are ineffective, as is synthetic phosphatidylcholine with two saturated alkyl chains. There appears to be an absolute requirement both for the presence of unsaturated alkyl chains and for the phosphatidyl choline head group. The former perhaps suggests that formation of a bilayer structure with a fluid core is a requirement, but a site on the protein specific for an unsaturated fatty acid chain is also possible. The specific head group requirement cannot be readily explained in nonspecific terms, nor does there seem any way in which the lipid head group can be involved directly in the enzymatic reaction. The inference is that the association between native enzyme and lipid involves a binding site with specificity for the head group. Although detergents (and presumably other lipids) can bind to the enzyme protein, they evidently leave the protein in an inactive conformation.

In other systems investigated by this technique specificity for cardiolipin and for phosphatidylglycerol has been observed. In some, such as the galactosyl transferase discussed below in relation to monolayer experiments, phospholipid is required without severe restriction on the nature of the head group. A general review is provided by Rothfield and Romeo (1971).

Experimental data of this type are less than satisfactory from the point of view of this chapter because they do not provide quantitative thermodynamic data. Moreover, because the phospholipid used in the experiments is added in aggregated form, they cannot distinguish between two very different mechanisms by which association between lipid and enzyme protein can occur. One possibility is that single lipid molecules bind to specific binding sites of the protein, and one may speculate that such binding involves only part of the lipid molecule, leaving another portion (perhaps one of the alkyl chains) free to associate with other lipid molecules and thereby assure incorporation in the membrane in the native situation or in the lipid aggregate in reactivation experiments. The second possibility is that a preformed aggregate of lipid is necessary before binding can occur. Studies at lipid concentrations below the cmc are necessary to distinguish between these alternatives.

MONOLAYER STUDIES

Lipid monolayers provide a useful tool for the study of protein–lipid interactions (Dawson, 1968). Protein can be introduced into the aqueous phase and its incorporation into the monolayer can be detected by an increase in surface pressure if the surface area is kept constant, or by an increase in area if the pressure is kept constant. Few systematic studies using this technique have been reported. One interesting study (Romeo et al., 1970a, b) involves a bacterial galactosyltransferase, which catalyzes the transfer of galactose from UDP-galactose to a lipopolysaccharide present in the bacterial cell envelope. An active enzyme system could be obtained by adding the inactive enzyme protein to the aqueous medium in contact with a mixed film of phosphatidylethanolamine and lipopolysaccharide at an air–water interface. A striking feature of the experiment, which was conducted at constant surface area, was a large increase in surface pressure, indicating penetration of the protein into the surface film. Earlier studies of this system by the technique of enzyme reactivation in the presence of phospholipid suspensions (Rothfield and Perlman, 1966) had shown a requirement for unsaturated acyl chains, but no striking head group specificity: negative results were obtained with phosphatidylcholine, but a variety of other head groups proved effective. It is thus likely that incorporation of the enzyme into the monolayer indicates the existence of a hydrophobic site capable of association with hydrocarbon chains. Without further work it is again impossible to conclude whether the process involves discrete specific sites on the protein (with specificity, say, for single hydrocarbon chains) or whether lipid in aggregated form is required.

LOCALIZATION OF THE SITE FOR ASSOCIATION WITH LIPIDS

Structural information has been obtained for two proteins that appear to be associated with membranes by virtue of possessing hydrophobic surface sites. Quantitative binding studies for lipids have not been carried out, but the structural data per se indicate that the sites for association with lipid are localized in each protein to a small portion of the polypeptide chain. It is surmised that only these portions of the protein are actually intimately associated with the cell membrane in vivo, and that the rest of the protein in each case extends into the adjacent aqueous medium.

One of these proteins is cytochrome b_5, a component of the endoplasmic reticulum of mammalian cells. It contains a noncovalently attached heme group, and it is functionally associated with a specific enzyme, cytochrome b_5 reductase, which catalyzes the transfer of electrons from DPNH to the heme iron atom. This protein can be isolated in soluble form with the aid of proteolytic enzymes, and with retention of its native heme linkage and native oxidation-reduction properties. The amino acid sequence of cytochrome b_5 isolated in this way from calf liver has been determined (Ozols and Strittmatter, 1969), and the three-dimensional structure has been obtained from X-ray diffraction studies of the crystallized protein (Mathews et al., 1971). By all criteria, the protein is a typical water-soluble protein. The protein has, however, been more recently isolated by use of detergents (Ito and Sato, 1968) and, when so prepared, has a larger molecular weight than when prepared by proteolysis and is no longer water-soluble. The increased molecular weight is contained in a segment of 44 amino acid residues with an unusually high proportion of hydrophobic amino acids (Spatz and Strittmatter, 1971). The results suggest that native cytochrome b_5 contains two principal domains. One is a globular domain with a highly polar surface, which contains the active site of the molecule and lies wholly outside the membrane proper. The other principal domain is smaller and it anchors the protein to the lipids of the membrane. The two domains are joined by a segment of extended polypeptide chain containing peptide bonds susceptible to the action of the proteolytic enzymes. The intact protein molecule, but not the proteolytic fragment, can be readily incorporated in vesicular preparations of microsomal lipids (Strittmatter et al., 1972).

The second protein that may have a highly localized site for interaction with lipids is the principal glycoprotein from erythrocyte membranes. This protein consists of a polypeptide chain of molecular weight about 22,000 and attached carbohydrate of molecular weight about 35,000. The carbohydrate is in the form of several distinct polysaccharide chains, attached at several positions on the amino terminal half of the polypeptide chain. Because of the hydrophilic nature of the carbohydrate moieties, this portion of the polypeptide chain must be located largely outside the membrane proper and, since the carbohydrate portions include blood group specific determinants and other recognition sites, it must be located outside the *external* membrane surface, and experimental studies confirm this (Marchesi et al., 1972). A partial amino-acid sequence of this protein has been obtained (Segrest et al., 1972), and has been found to contain a stretch of 23 residues with a remarkably high proportion of hydrophobic amino acids and containing not a single charged residue. If this stretch of amino acids were in an α-helical conformation, it would form a rod with a hydrophobic surface and with a length (35 Å) comparable to the thickness typical of the hydrophobic core of a lipid bilayer (Fig. 12-2). Moreover, immediately following this stretch of 23 residues (toward the carboxy terminal end of the chain) is a sequence of six residues containing four positively charged side chains, ideally suited for association with lipids containing negatively charged head groups. It has been suggested (Bretscher, 1971; Segrest et al., 1972) that this protein spans an erythrocyte membrane from one side to the other, and that all parts other than the 29 amino-acid residues specifically mentioned here exist in vivo in the aqueous media outside and inside the cell.

CHOLESTEROL AND RELATED COMPOUNDS

The biological significance of the presence of cholesterol in some membranes is not known, and one possibility is that it may be involved in the association of proteins with membranes. Thus studies of the association between proteins and cholesterol and other molecules containing the steroid ring are of considerable interest.

It has been found (Makino and Tanford, 1973) that the bile salt anion, deoxycholate, binds to the amphiphile binding sites of native serum albumin with an affinity roughly equal to that of decanoate, that is, a binding constant of the order of 10^5 liters/mole. Since much of the free energy of association at these sites can be ascribed to the COO^- head group of deoxycholate, it is evident that the hydrophobic portions of these sites have no special affinity for the steroid ring. Numerous steroids lacking an anionic group have however been found to associate with serum albumin, albeit quite weakly: most of them have a binding constant of order 10^4 liters/mole. (See review by Steinhardt and Reynolds, 1969). Weak binding is observable for these substances because they possess more than one hydrophilic group, at widely separated points of the steroid ring, which presumably greatly weakens the tendency for competitive self-association processes.

On the other hand, no significant association between cholesterol and serum albumin takes place (J. A. Reynolds, unpublished results). The reason for this is undoubtedly the remarkably strong tendency for self-association of cholesterol in water (page 106), which limits the concentration of monomeric cholesterol in water to about $10^{-8}M$. Thus a binding constant of the order of 10^8 liters/mole would be required for experimentally observable association. This is several orders of magnitude greater than the association constants measured for other steroids. Whether proteins other than serum albumin have the requisite affinity is not known at this time.

Cholesterol is very soluble in organic solvents, and it was shown in Chapter 12 that it associates readily with phospholipid bilayers or monolayers, with the hydroxyl group at the hydrophilic surface and the rest of the molecule intercalated between the hydrocarbon chains of the phospholipid. Cholesterol may be able to associate in a similar fashion with protein–amphiphile complexes, but it appears unable to do so with the micellar protein–dodecyl sulfate complex described in the previous chapter, perhaps because the hydrophobic layer in this complex is not sufficiently deep.

REFERENCES

Bretscher, M. S. (1971). *Nature New Biol.*, **231,** 225.

Das, M. L., and F. L. Crane. (1964). *Biochemistry*, **3,** 696.

Das, M. L., E. D. Haak, and F. L. Crane. (1965). *Biochemistry*, **4,** 859.

Dawson, R. M. C. (1968). In *Biological Membranes*, D. Chapman, Ed., Academic Press, New York, Chapter 5.

Fleischer, B., A. Casu, and S. Fleischer. (1966). *Biochem. Biophys. Res. Comm.*, **24,** 189.

Gulik-Krzywicki, T., E. Shechter, V. Luzzati, and M. Faure. (1969). *Nature*, **223,** 1116.

Ito, A., and R. Sato. (1968). J. Biol. Chem., **243,** 4922.

Ji, T. H., and A. A. Benson. (1968). *Biochim. Biophys. Acta*, **150,** 686.

Kimelberg, H. K., and D. Papahdjopoulos. (1971). *J. Biol. Chem.*, **246,** 1142.

Kimelberg, H. K., C. P. Lee, A. Claude, and E. Mrena. (1970). *J. Membrane Biol.*, **2,** 235.

Makino, S., and C. Tanford. (1973). *Fed. Proc.*, **32,** 572 Abs.

Marchesi, V. T., T. W. Tillack, R. L. Jackson, J. P. Segrest, and R. E. Scott. (1972). *Proc. Nat. Acad. Sci. U.S.A.*, **69,** 1445.

Mathews, F. S., M. Levine, and P. Argos. (1971). *Nature New Biol.*, **233,** 15.

Ozols, J., and P. Strittmatter. (1969). *J. Biol. Chem.*, **244,** 6617.

Palmer, F. B., and R. M. C. Dawson. (1969). *Biochem. J.*, **111,** 637.

Papahdjopoulos, D., and J. C. Watkins. (1967). *Biochim. Biophys. Acta*, **135,** 639.

Reynolds, J. A., and C. Tanford. (1970). *Proc. Nat. Acad. Sci U.S.A.*, **66,** 1002.

Romeo, D., A. Girard, and L. Rothfield. (1970a). *J. Mol. Biol.* **53,** 475.

Romeo, D., A. Hinckley, and L. Rothfield. (1970b). *J. Mol. Biol.*, **53,** 491

Rothfield, L., and M. Pearlman. (1966). *J. Biol. Chem.*, **241,** 1386.

Rothfield, L. and D. Romeo. (1971). In *Structure and Function of Biological Membranes*, L. Rothfield, Ed., Academic Press, New York, Chapter 6.

Segrest, J. P., R. L. Jackson, and V. T. Marchesi. (1972). *Biochem. Biophys. Res. Comm.*, **49,** 964.

Sekuzu, I., P. J. Jurtshuk, and D. E. Green. (1963). *J. Biol. Chem.*, **238,** 975.

Shipley, G. G., R. B. Leslie, and D. Chapman. (1969a). *Nature*, **222,** 561.

Shipley, G. G., R. B. Leslie, and D. Chapman. (1969b). *Biochim. Biophys. Acta*, **173,** 1.

Spatz, L., and P. Strittmatter. (1971). *Proc. Nat. Acad. Sci. U.S.A.*, **68,** 1042.

Steinemann, A., and P. Läuger. (1971). *J. Membrane Biol.*, **4,** 74.

Steinhardt, J., and J. A. Reynolds. (1969). *Multiple Equilibria in Proteins*, Academic Press, New York.

Strittmatter, P., M. J. Rogers, and L. Spatz. (1972). *J. Biol. Chem.*, **247,** 7188.

Sweet, C., and J. E. Zull. (1970). *Biochim. Biophys. Acta*, **219,** 253.

Taborsky, G. (1970). *Biochemistry*, **9,** 3768.

SEVENTEEN

SERUM LIPOPROTEINS

Soluble lipoproteins are not only of interest for their own sake, but are also an essential topic for a discussion of biological membranes. An understanding of the interaction of lipids with proteins in the formation of membranes would not be complete unless we also understand how a similar type of interaction can lead to the formation of soluble particles with no tendency for incorporation into membranes. This chapter will summarize the present state of knowledge about the best studied proteins of this kind, the serum lipoproteins. Lipoproteins from egg yolk have also received a fair amount of study, and a summary of their properties may be found in a review by Cook and Martin (1969).

Because of their high lipid content, the serum lipoproteins have a lower buoyant density than other serum proteins, and the classical method for their isolation, by flotation in salt solutions of different density, is based on this. Four major "fractions" are obtained in this way (Nichols, 1969), and at least three of them contain major components that are distinct molecular entities, with unique polypeptide chains and unique functional properties. Allowing for the possibility that a distinct protein may also emerge from the fourth fraction (the chylomicrons), four different serum lipoproteins may exist, as follows.

Chylomicrons

This fraction of lowest buoyant density consists of very large particles. Their function is to transport fats derived from exogenous food sources to fat storage depots. They typically contain less than 2% protein, and about 90% of storage fats in the form of triglycerides (i.e., triacyl derivatives of glycerol).

The precise compositon depends on the diet of the individual. Since triglycerides themselves have extremely low solubility in aqueous solution, the mechanism whereby they are solubilized by so small an amount of protein and other lipids represents an intriguing problem that is at present completely unsolved. No unique polypeptide chains have been identified with this fraction, and it is possible that the chylomicrons are simply a very triglyceride-rich form of VLDL (Kostner and Holasek, 1972). For a review, see Zilversmit (1969).

Very Low Density Lipoproteins (VLDL)

VLDL again represents a "fraction" instead of a purified protein. However, three polypeptide chains have been uniquely identified with this fraction (Brown et al., 1970; Herbert et al., 1971). They are all of low molecular weight (7000 to 10,000). Particles isolated from this fraction have molecular weights between 5 and 10×10^6 (Levy et al., 1971), and contain about 8% protein, which corresponds to 50 or more polypeptide chains per particle. VLDL, like the chylomicrons, is involved in the transport of triglycerides, primarily from fat storage depots for use elsewhere in the body. VLDL particles vary in their lipid content and, typically, contain 50% triglycerides by weight.

Low Density Lipoprotein (LDL)

This is the most abundant serum lipoprotein and a reasonably well-characterized molecular entity. It contains unique polypeptide chains of molecular weight 255,000 (Smith et al., 1972). The lipoprotein, as normally isolated from the appropriate buoyant density fraction, has a molecular weight of about 2.6×10^6 (Adams and Schumaker, 1969; Fisher et al., 1971) and a protein content of about 20% by weight, that is, there are two polypeptide chains per molecule. The same protein, as isolated by alcohol fractionation of the serum proteins, has a molecular weight of 1.3×10^6 (Oncley et al., 1947), and there is other evidence to indicate that the protein as normally isolated may be a readily dissociable dimer. The lipid composition is somewhat variable, but a unique feature is a high content of cholesteryl esters (37% by weight). The function of the cholesteryl esters is not known, but it has been established that the esterified fatty acid chains are readily transferred from cholesteryl esters to lysophosphatidyl cholines to form diacyl phosphatidyl cholines, and vice versa.

High Density Lipoprotein (HDL)

This is another reasonably well-characterized molecular entity. It contains two distinct polypeptide chains, designated AI and AII. AI has a molecular weight of about 28 to 29,000; AII has a molecular weight of 8700, and its amino-acid sequence has been determined (Brewer et al., 1972). AII is present in the protein as a disulfide-bonded dimer. The HDL lipoprotein molecule is capable of existing in a variety of forms, varying in total molecular weight (from about 175,000 to 400,000 or more), and in the relative amounts of protein and lipid. Both polypeptide chains are always present, and the total protein content is always larger than for other lipoproteins. The phospholipid content of HDL is always large, and one of its functions may be to provide a reservoir of phospholipid for incorporation into membranes. HDL has a high content of cholesteryl esters, and is probably the primary site for exchange of acyl chains between cholesterol and phosphatidyl choline, even though a much larger amount of the total cholesteryl ester of serum is carried by LDL.

The variable lipid content of the serum lipoproteins, a necessary attribute of their function as transport proteins, generally makes it impossible to achieve perfect separation on the basis of buoyant density alone. Thus the "fractions" from which given lipoproteins are isolated are generally found to contain, on the basis of polypeptide chain analysis or by immunochemical tests, small amounts of the polypeptide chains of the other proteins. There is growing evidence, however, (Albers et al., 1972) that association between polypeptide chains of the different lipoproteins to form hybrid *molecules* does not take place.

Some of the molecular properties of VLDL, LDL, and HDL are summarized in Table 17-1. The lipid analyses given in this table refer to the appropriate buoyant density "fractions" and so do not necessarily reflect the compositions of the purified proteins accurately. For example, it is not known whether the triglycerides are constitutive components of LDL and HDL, or nonspecifically dissolved in the constitutive lipid (see Chapter 10), or (at least in part) artifacts arising from contamination of the sample used for analysis by VLDL. An interesting feature of the lipid content of all of the lipoproteins is the preponderance of phosphatidylcholine and sphingomyelin among the phospholipids. Moreover, glycolipids are present in only trace amounts. The lipid compositions thus differ from those of cell membranes (Table 12-1) not only by virtue of the presence of triglycerides and cholesteryl esters, but also in the relative amounts of the lipid classes that they have in common with cell membranes. These differences exist even though exchange of lipids between serum lipoproteins and membranes (notably erythrocyte membranes) takes place freely.

Table 17-1. Molecular Properties of Serum Lipoproteins[a]

	VLDL[b]	LDL[c]	HDL[d]	
Polypeptide chains with molecular weight	D^1 7,000 D^2 10,000 $D_{3,4}$ 10,000	255,000	AI 28,500 AII 8,700	
			HDL_2	HDL_3
Total molecular weight	5–10 × 10^6	2.6 × 10^6	360,000	175,000
% Protein	~8	20	40	55
Number of chains per molecule	⪆50	2	3 AII 8 AII	2 AI 4 AII
Nonprotein moiety[e]				
% triglycerides	65	13	8	10
% cholesteryl esters	5	48	29	28
% free cholesterol	10	10	10	8
% phospholipid[f]	20	29	53	54

[a] These data are for the human lipoproteins, but similar proteins are probably present in all mammals. The data refer to the *major* species containing the polypeptide chains at the head of each column. When these same polypeptide chains occur in other than the normal buoyant density fraction, the overall protein molecular weight and composition will probably be different.

[b] The polypeptide chain molecular weights of VLDL were obtained by procedures deemed unreliable by the author and are subject to an error of perhaps ±25%. Polypeptides D_3 and D_4 are identical in amino-acid composition. They contain small amounts of carbohydrate and differ in their sialic-acid content.

[c] There is evidence (see text) that LDL may also exist with a molecular weight of 1.3 × 10^6, containing a single polypeptide chain.

[d] HDL_2 and HDL_3 are two particular species that have been studied in detail (Hazelwood, 1958; Scanu, 1969, 1971). Other species of somewhat different composition probably exist. The polypeptide chain compositions of HDL_2 and HDL_3 were estimated by the author on the basis of various analytical data. An even number of AII chains per molecule is required because AII is present as a disulfide-bonded dimer.

[e] Lipid analyses refer to the buoyant density "fractions" from which the proteins are derived and may differ from the lipid compositions of pure proteins. The figure for triglycerides includes small amounts of incompletely esterfied glycerol (diglycerides, monoglycerides). Small amounts of free fatty acid, hydrocarbon and other presumably "dissolved" substances have been omitted from the tabulation. The data for VLDL and LDL are from Levy et al. (1971), those for HDL from Scanu (1971). Much more detailed analyses are given by Skipsky et al. (1967).

[f] 90% of the phospholipid consists of phosphatidylcholine and sphingomyelin. Both these lipids contain the zwitterionic choline head group. Note also that glycolipids are present in only trace amounts.

An interesting property of the polypeptide chains of all of the serum lipoproteins is that they do not have unusual overall amino-acid compositions; that is, they resemble water-soluble proteins, containing about 25 to 30% strongly hydrophobic residues and about 45 to 50% ionic and strongly polar residues. This eliminates the possibility that the ability of these polypeptide chains to combine with large amounts of lipid arises from an unusually large content of hydrophobic amino acids, and consequent inability to form compact structures with a predominantly hydrophilic surface. In the one complete amino-acid sequence that has been determined, that of the AII chain of HDL (Brewer et al., 1972), nothing unusual can be seen in the distribution of polar and apolar amino acids along the chain, so that at least in this case there is no indication that even a part of the polypeptide chain might be specially adapted for formation of a highly hydrophobic surface.

STRUCTURAL STUDIES

The approximate shapes of LDL and HDL molecules were first determined by Oncley et al. (1947) on the basis of measurements of the intrinsic viscosity and frictional coefficient. LDL was found to be compact, nearly spherical, resembling a typical globular protein, but the data indicated that HDL was less symmetrical. Subsequent work has confirmed the earlier result for LDL (Fisher et al., 1971), but not for HDL, and more recent measurements indicate that it, too, forms compact globular particles. The intrinsic viscosity, for example, is between 3.2 and 3.5 cc/g for both HDL_2 and HDL_3 (Scanu, 1969). Electron micrographs indicate that chylomicrons and VLDL, as well as LDL and HDL, exist as compact, nearly spherical particles (e.g., Nichols, 1969). Some of the electron micrographs for LDL and HDL suggest the existence of small subunits within the compact particles, but insufficient work has been done to assess the possible significance of this observation.

Considerable effort has been expended on attempts to obtain more detailed structural information for LDL and HDL, but so far without much success. The arrangement of protein and lipid within these and other lipoprotein molecules is still largely unknown, although some aspects of this arrangement can be limited to a few possibilities on the basis of the available data and the known properties of the lipid constituents. No information of any kind is available on the linkage between protein and lipid. The work done on LDL has been summarized by Margolis (1969) and that on HDL by Leslie (1971).

The major problems in lipoprotein structure will be discussed here in a manner different from that in existing reviews. We shall begin by considering each distinct chemical component alone, and try to arrive at reasonable conclusions concerning its possible structural state in a lipoprotein molecule. The linkage between protein and lipid, which is the most important aspect of the overall problem, will be discussed only at the end of the chapter.

(1) Triglycerides

The ester bond in triglycerides is not significantly hydrophilic, and these substances do not form soluble micelles in water. They can of course enter into micelles formed by other amphiphiles (as cholesterol, for instance, has been shown to do). In chylomicrons and VLDL, however, the triglycerides constitute such a large fraction of the total mass, that they must represent the major structural element. The chylomicron and VLDL particles must be essentially globules of liquid fat, that are somehow kept in solution or suspension in water by the protein and phospholipid with which they are combined. It is logical to imagine a hydrophilic surface film over the fat glubule, and there is evidence for this from electron microscopy (Zilversmit, 1969). Even in VLDL, however, there is not enough protein and phospholipid to cover the surface of the triglyceride core completely (Margolis, 1969), and in chylomicrons, with a 90% content of triglycerides, only a fraction of the surface can conceivably be covered (Zilversmit, 1969). Since our main interest here is in the structure of LDL and HDL, we shall not pursue this problem further. It is important, however, in considering possible structures for LDL and HDL, to keep in mind the possibility of some analogy with the chylomicron and VLDL structures.

(2) Cholesteryl Esters

The cholesteryl esters, like the triglycerides, are soluble in organic solvents and do not possess sufficiently hydrophilic groups to form micelles in water. When present in HDL and LDL they, too, may be in the form of liquid droplets in the molecular core. There is evidence to the contrary, however, suggesting that the cholesteryl esters play a specific structural role, at least in HDL. Forte and Nichols (1971) have found that HDL molecules reconstituted from the constituent polypeptide chains, phosphatidylcholine and free cholesterol appear as stacked, disclike structures in electron micrographs. When cholesteryl esters were added, molecules more nearly resembling normal HDL were obtained. Disclike structures were also observed in the HDL from abnormal individuals who are deficient in the enzyme for transferring acyl chains from phosphatidylcholine to cholesterol. It has also been reported (Lux et al., 1972) that the circular dichroism spectrum of native HDL can be restored by combining the delipidated polypeptide chains with phosphatidylcholine and cholesteryl oleate, whereas restoration is only partial if the cholesteryl oleate is omitted. The changes in the circular dichroism spectrum that are involved here are small, however, and as will be discussed below, argue against massive interaction between the polypeptide chains of HDL and any of the constituent lipid. Thus, if cholesteryl esters do play an important structural role, it is likely that interactions with other lipids are involved. Such interactions need not be *specific* to produce the effects observed by electron microscopy: for example, the cholesteryl esters in native HDL could be in the form of a droplet covered by a phospholipid monolayer. In their absence, the monolayer could coalesce to form a bilayer and the consequent spatial rearrangement might account for the electron microscopy results.

(3) Free Cholesterol

As was pointed out in Chapter 12, the solubility of cholesterol in water, even in the form of aggregates, is extremely small. On the other hand, it is readily incorporated into phospholipid monolayers or bilayers, and it would also be readily soluble in oil droplets, if such exist as part of the overall structure. The maximum amount of cholesterol that can be incorporated into a phospholipid monolayer or bilayer is one mole per mole of phospholipid, and the cholesterol content of none of the serum lipoproteins exceeds this amount. Thus one possibility is that the type of structure shown in Fig. 12-3, or the analogous monolayer structure, is present in the lipoproteins also.

(4) Phospholipids

It has been seen in previous chapters that diacyl phospholipids alone, or in combination with free cholesterol, invariably aggregate to form bilayers. Small vesicles, bounded by a bilayer and containing a water-filled cavity, are readily formed. However, by sonication of phospholipid in the presence of organic solvent, vesicles with an organic solvent core, bounded by a monolayer, can also be formed (page 110). Thus the phospholipid in serum lipoproteins could be present in the form of a bilayer, or as a monolayer around fat or cholesteryl ester droplets (if they are present). The major components among the phospholipids are phosphatidylcholine and sphingomyelin, and the fatty acid chains attached to them contain about 50% unsaturated paraffin chains (Goodman and Shiratori, 1964). If bilayers or monolayers are present, they will therefore have liquid cores, with the structural features discussed in Chapters 12 and 13. Some of the phospholipid, perhaps a major part of it, could of course be involved in interactions with the polypeptide chains, forming structures unlike any that are found in the absence of protein.

High resolution pmr measurements have been made with both LDL and HDL (Steim et al., 1968; Chapman et al., 1969). Sharp resonance peaks are obtained for the protons from hydrocarbon CH_2 and CH_3 groups, as well as for those from the methyl groups of the choline head groups. The patterns obtained are indistinguishable from those obtained when the lipids extracted from the lipoproteins are dispersed in water. There can be no doubt that the bulk of the phospholipid is in a fluid environment such as would obtain in bilayer or monolayer arrangements unperturbed by interaction with protein. It is significant that the same result is not obtained for erythrocyte membranes, where the presence of protein clearly has some restraining influence on the motion of the lipid. A result such as this must be interpreted with caution, since the resonance peaks do not necessarily represent all of the lipid molecules. Quantitation of the areas under the peaks indicates however, that at least 90 to 95% of the constituent phospholipid contributes to the observed spectra. Thus it does not seem possible that more than 10% of the phospholipid alkyl chains or head groups can be involved in interactions with the protein moieties of HDL or LDL.

Corroboration for the location of most of the phospholipid *head groups* at the molecular surface is provided by the accessibility of 80% or more of these groups, both in LDL and HDL, to the action of phospholipases, enzymes that specifically cleave the lipid molecules at various locations on the head group (Margolis, 1969; Leslie, 1971). These experiments may also indicate that the bulk of the phospholipid is not present as a vesicle bounded by a bilayer, since the internal surface of a bilayer would be inaccessible to the enzymes.

(5) Protein Moiety

The identification of unique polypeptide chains as belonging to LDL and HDL has occurred only recently and very few experiments focusing on the properties of the individual chains have been carried out. As mentioned earlier, their overall amino-acid compositions are similar to those of water-soluble proteins. The AI and AII chains of HDL are in fact water-soluble in the absence of lipid. There is evidence that a specific site for association between AI and the disulfide-bonded dimer of AII may exist (R. Simon and J. A. Reynolds, unpublished data), so that the soluble apoprotein of HDL may exist largely as a complex between them. A significant observation, first made by Scanu (1965), is that the optical rotatory properties of the apoprotein are very similar to those of native HDL. Recent precise measurements of the circular dichroism spectrum (Lux et al., 1972) show that there are differences in magnitude, but not in the general features of the spectrum in the 200 to 240 mμ region, which reflects the folding of the polypeptide backbone. Peaks at 258 and 264 mμ (probably associated with the local environment of phenylalanyl residues) also appear to be the same in apo-HDL and in the native lipoprotein. Marked effects of removal of the lipid appear only in the 290 mμ region, indicating that tryptophan side chains acquire an altered environment. Another interesting observation by Scanu et al. (1968) is that most of the lysyl amino groups of the protein are readily accessible for modification by chemical reagents in the intact native protein. Taken together, these results suggest that the protein moiety of HDL is a water-soluble complex with a structure like that of other water-soluble proteins and that interaction with lipid to form the lipoprotein involves only limited portions of the polypeptide chains.

Solubilization of the very large polypeptide chains of LDL in the absence of lipid is much more difficult (Scanu et al., 1969), and it has not been established that the intact protein moiety can be maintained in solution at all unless detergents are added to replace the lipid. Helenius and Simons (1971) have shown that deoxycholate or a neutral detergent can be used to obtain soluble apo-LDL with retention of the immunochemical determinants of native LDL, which perhaps suggests, but certainly does not prove, that the interaction between protein and lipid in native LDL, similar to that in native HDL, involves only limited portions of the polypeptide chain. More definite evidence about the existence of the protein moiety of LDL as a largely independent globular entity in the native lipoprotein comes from the early physical studies of Oncley et al. (1950). They showed that the pH-dependence of the solubility of LDL resembles that of a typical water-soluble protein and that most of the acidic and basic groups of the protein could be titrated without obvious indication of denaturation or separation of lipid from protein.

(6) Overall Structure and Linkage Between Lipid and Protein

The foregoing results favor overall structures for both HDL and LDL in which the lipid and protein form largely separate entities. The lipid may be in the form of globules of neutral lipids covered by a monolayer of phospholipid, but other possibilities are by no means excluded. The polypeptide chains would appear to be folded into one or more globular domains. Some short sequences of the polypeptide chains could consist of hydrophobic amino-acid residues that enter the interior of the lipid globule, but the interaction between protein and lipid could equally well reside at the surface of the folded protein domains. In the case of HDL one or more tryptophan residues may be intimately involved. Clearly much more work must be done before a plausible proposal about specific details of the linkage between protein and lipid can be made.

One type of model for lipoprotein structure that can be completely rejected with some confidence is a model consisting of a protein core completely surrounded by lipid.

REFERENCES

Adams, G. H., and V. N. Schumaker. (1969). *Anal. Biochem.*, **29**, 117.

Albers, J. J., C. H. Chen, and F. Aladjem. (1972). *Biochemistry*, **11**, 57.

Brewer, H. B., Jr., S. E. Lux, R. Ronan, and K. M. John. (1972). *Proc. Nat. Acad. Sci. U.S.A.*, **69**, 1304.

Brown, W. V., R. I. Levy, and D. S. Fredrickson. (1970). *J. Biol. Chem.*, **245**, 6588.

Chapman, D., R. B. Leslie, R. Hirz, and A. M. Scanu. (1969). *Biochim. Biophys. Acta*, **176**, 524.

Cook, W. H., and W. G. Martin. (1969). In *Structural and Functional Aspects of Lipoproteins in Living Systems*, E. Tria and A. M. Scanu, Eds., Academic Press, New York.

Fisher, W. R., M. E. Granade, and J. L. Mauldin. (1971). *Biochemistry*, **10**, 1622.

Forte, T., and A. V. Nichols. (1971). *Biophys. Soc. Abstr.*, 125a.

Goodman, D. S., and T. Shiratori (1964). *J. Lipid Res.*, **5**, 307.

Hazelwood, R. N. (1958). *J. Am. Chem. Soc.*, **80**, 2152.

Helenius, A., and K. Simons. (1971). *Biochemistry*, **10**, 2542.

Herbert, P., R. I. Levy, and D. S. Fredrickson. (1971). *J. Biol. Chem.* **246**, 7068.

Kostner, G., and A. Holasek. (1972). *Biochemistry*, **11**, 1217.

Leslie, R. B. (1971). In *Plasma Lipoproteins*, R. M. S. Smellie, Ed., Biochemical Society Symposium No. 33, Academic Press, New York, p. 47.

Levy, R. I., D. W. Bilheimer, and S. Eisenberg. (1971). In *Plasma Lipoproteins*, R. M. S. Smellie, Ed., Biochemical Society Symposium No. 33, Academic Press, New York, p. 3.

Lux, S. E., R. Hirz, R. I. Shrager, and A. M. Gotto. (1972). *J. Biol. Chem.*, **247**, 2598.

Margolis, S. (1969). In *Structural and Functional Aspects of Lipoproteins in Living Systems* E. Tria and A. M. Scanu, Eds., Academic Press, New York.

Nichols, A. V. (1969). *Proc. Nat. Acad. Sci. U.S.A.*, **64**, 1128.

Oncley, J. L., F. R. N. Gurd, and M. Melin. (1950). *J. Am. Chem. Soc.*, **72**, 458.

Oncley, J. L., G. Scatchard, and A. Brown. (1947). *J. Phys. and Colloid Chem.*, **51**, 184.

Scanu, A. M. (1965). *Proc. Nat. Acad. Sci. U.S.A.*, **54**, 1699.

Scanu, A. (1969). In *Structural and Functional Aspects of Lipoproteins in Living Systems*, E. Tria and A. M. Scanu, Eds., Academic Press, New York.

Scanu, A. M. (1971). In *Plasma Lipoproteins*, R. M. S. Smellie, Ed., Biochemical Society Symposium No. 33. Academic Press, New York, p. 29.

Scanu, A. M., W. Reader, and C. Edelstein. (1968). *Biochim. Biophys. Acta*, **160**, 32.

Scanu, A. N., H. Pollard, R. Hirz, and K. Kothary. (1969). *Proc. Nat. Acad. Sci. U.S.A.*, **62,** 171.
Skipsky, V. P., M. Barclay, R. K., Barclay, V. A. Fitzer, J. J. Good, and F. M. Archibald. (1967). *Biochem. J.*, **104,** 340.
Smith, R., J. R. Dawson, and C. Tanford. (1972). *J. Biol. Chem.*, **247,** 3367.
Steim, J. M., O. J. Edner, and F. G. Bargoot. (1968). *Science*, **162,** 909.
Zilversmit, D. B. (1969). In *Structural and Functional Aspects of Lipoproteins in Living Systems*, E. Tria and M. Scanu, Eds., Academic Press, New York, Chapter C1.

EIGHTEEN

BIOLOGICAL MEMBRANES AND THEIR COMPOSITION

Biological membranes are thin layers of protein and lipid that permit the compartmentalization of living matter. The simplest example is the cytoplasmic membrane (also called "plasma membrane") that separates the contents of a single cell from its environment. An example is shown in Fig. 18-1. Simple membranes of this type surround all living cells. They serve as barriers to prevent the mixing of the contents of the cell with its surroundings, but at the same time must permit the selective passage of metabolites in and out of the cell. In bacteria the cell membrane is usually (but not always) surrounded by a second barrier, called the *cell wall.*

The cells of higher organisms (eukaryotic cells) are much larger than the cells of bacteria and similar organisms (prokaryotic cells). They usually contain intracellular membranes that define compartments within the cell. The cell nucleus and mitochondria are examples, each being surrounded by its own membrane. Eukaryotic cells also contain endoplasmic reticulum, which is a network of membrane-bounded channels that traverse the entire cell. The mature human erythrocyte (Fig. 18-1) is thus an atypically simple type of eukaryotic cell, since it contains no nucleus or other intracellular structures.

In mitochondria there are actually two membranes, as illustrated by Fig. 18-2. The inner membrane contains many inward folds entering into the body of the mitochondrion.

Higher organisms contain specialized membranes. A well-known example is the myelin sheath surrounding the axon of a nerve cell, as shown in Fig. 18-3. This structure is formed by extension of the cytoplasmic membrane of a cell called the *Schwann cell* (a glial cell is used by peripheral nerves), which

Fig. 18-1. Electron micrograph of a cross section through a mature erythrocyte, showing the cytoplasmic membrane surrounding the cell. The cell has been stained with osmium tetroxide and the membrane appears as two electron dense lines with an intervening space (provided through the courtesy of Dr. J. D. Robertson).

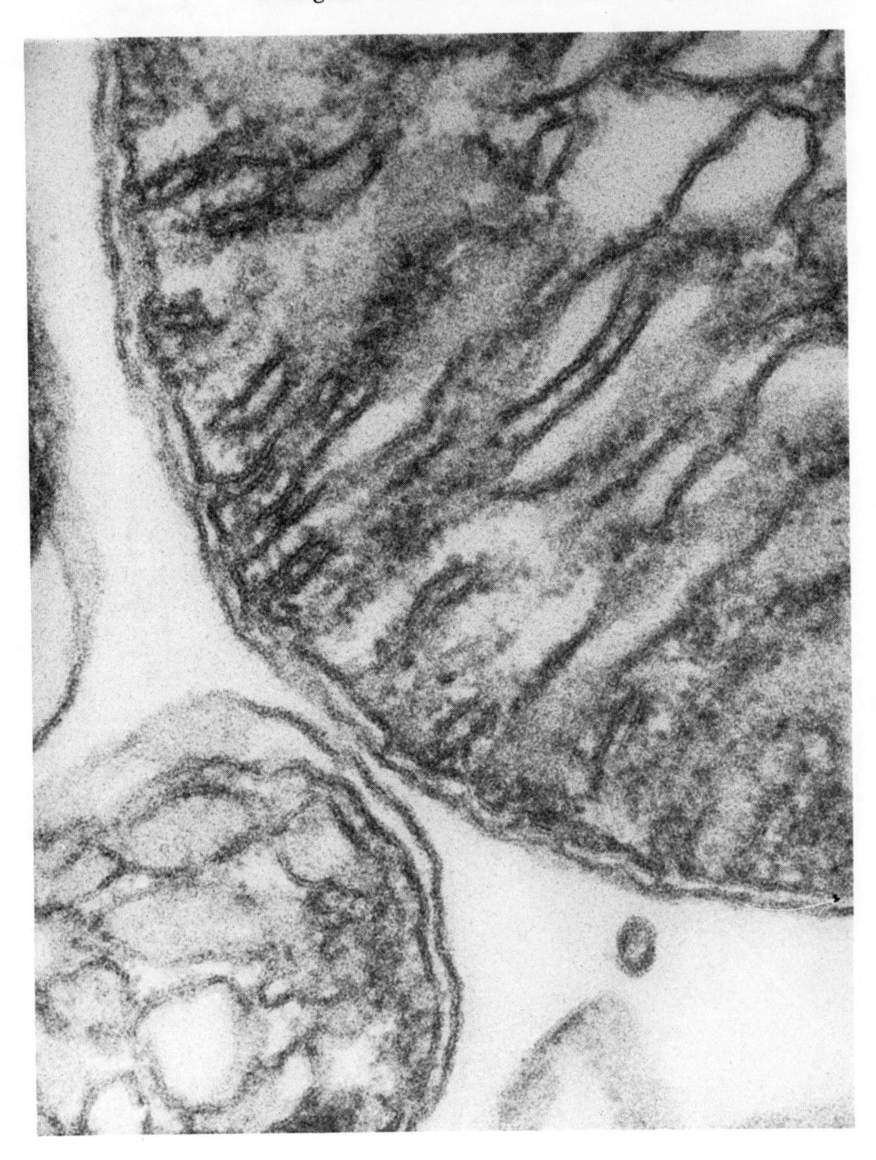

Fig. 18-2. Electron micrograph of a cross section through a mitochondrion. The mitochondrion has a smooth outer membrane and a separate inner membrane with many folds into the interior of the organelle. (Stoeckenius, 1969). From *Membranes of Mitchondria and Chloroplast* by Prof. E. Racker © 1970 by Litton Educational Publishing, Inc. Reprinted by permission of Van Nostrand Reinhold Company.

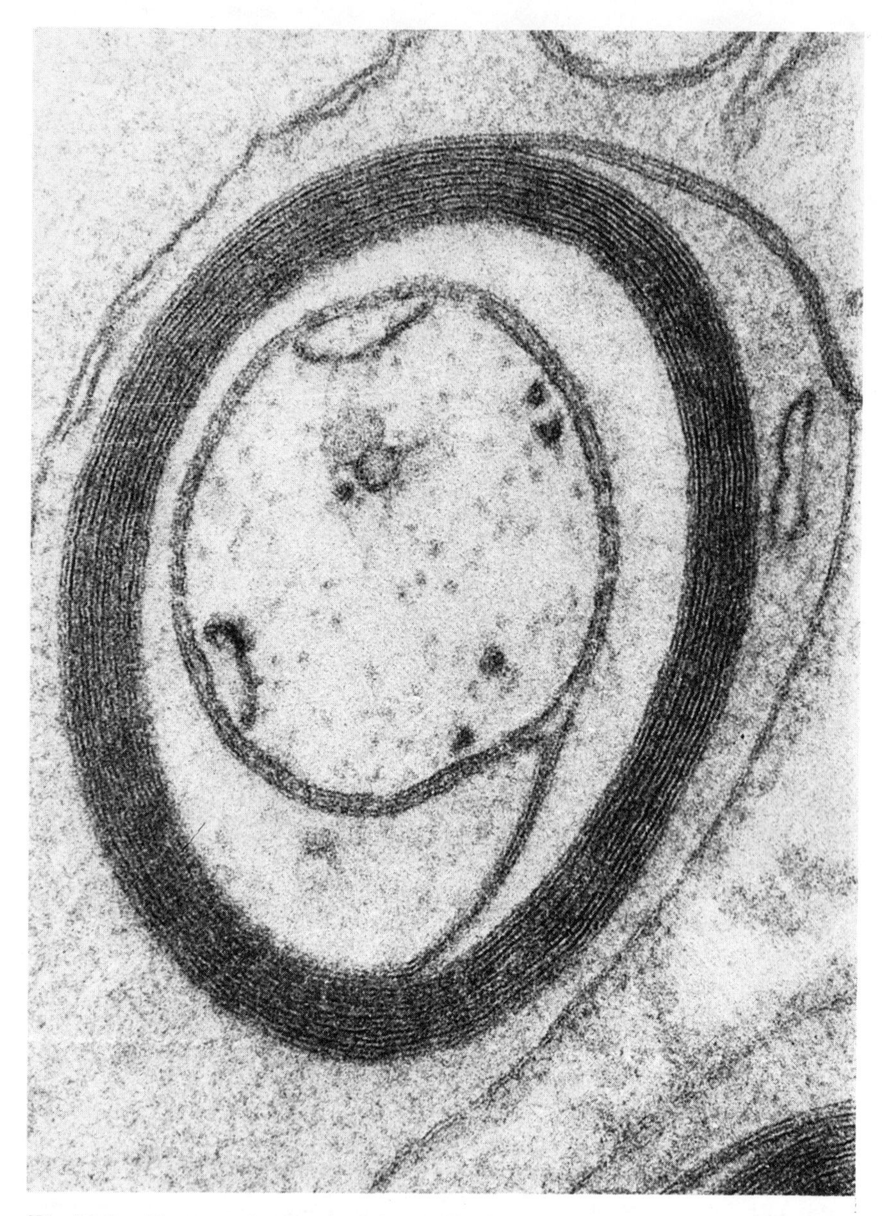

Fig. 18-3. Electron micrograph of the myelin sheath around a nerve axon. The innermost membrane is the axonal membrane, and the surrounding sheath is derived from the plasma membrane of a Schwann cell (provided through the courtesy of Dr. J. D. Robertson).

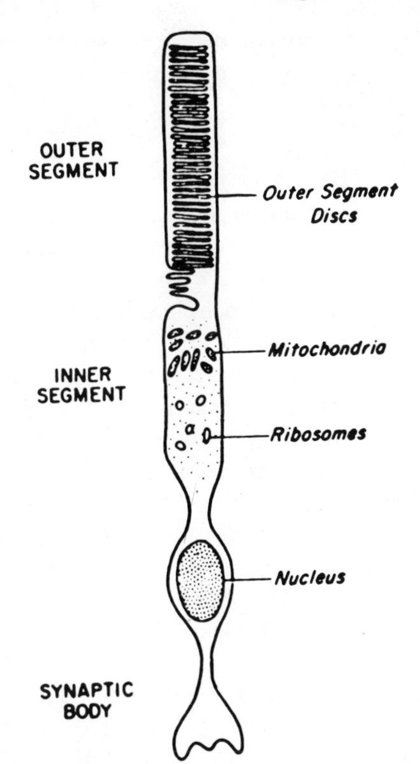

Fig. 18-4. Schematic drawing of a rod photoreceptor cell. The outer segment discs are formed by inward folding of the cytoplasmic membrane, followed by detachment (taken from Heller and Ostwald, 1972).

wraps itself in several layers about the axonal membrane. Another example is provided by the stacked discs of the outer segments of retinal rod cells, illustrated by Fig. 18-4. These discs are formed by folding of the outer membrane of the cell and eventually disconnected from it. A similar folding of the outer membrane occurs in retinal cone cells, but the folds do not become disconnected into discrete discs.

We should note finally that certain viruses have membranous outer shells. They are derived from the cytoplasmic membranes of the host cell, which encapsulates the virus as it leaves the cell. The proteins of the host membranes appear to be generally lost in this process.

All of these different types of membranes share an important characteristic: they contain proteins and lipids and no other major constituents. They contain a small amount of carbohydrate in combined form, as glycolipid (page 96) or as glycoproteins, but no free polysaccharide. Bacterial cell walls, on the other hand, contain a rigid covalently linked polysaccharide–peptide complex as a major constituent. They are sometimes called "outer membranes," but this nomenclature is not recommended because it obscures the fundamental compositional and structural difference between the cell wall and all other types of membranes here under discussion.[1]

OVERALL COMPOSITION

Accurate knowledge of the composition of membranes is not yet available. In part this is due to the difficulty in isolating membranes free of cytoplasm and other components of the organism. Thus bacterial cytoplasmic membranes may be contaminated by cell wall constituents, endoplasmic reticulum may be contaminated by attached ribosomes, and so on. Purification is achieved most easily in systems in which the desired membrane is present in large amount and other membranous structures are absent. Erythrocyte membranes fall into this category, as do the membranes of microorganisms not containing cell walls (e.g., *mycoplasma*) and viral membranes. Retinal rod membranes can also be obtained in a high state of purity. Myelin is almost pure as it exists in nature since the amount of myelin membrane greatly exceeds the amount of cytoplasmic membrane of the Schwann or glial cell from which it is derived. It has not been established whether preparations can be obtained entirely free from the axonal membrane.

Even when membranes can be isolated free from contaminants, a minor ambiguity about the composition always remains, since loosely held components may be removed in the process of washing the membrane free from cytoplasm and intercellular fluid.

Some analytical data on the overall compositions of typical membranes are given in Table 18-1, and they should be considered with some reservation in view of the foregoing uncertainties. It seems clear, however, that most membranes contain by analytical test at least as much protein as lipid, and that a considerable excess of protein over lipid is not uncommon. Myelin membrane is exceptional among all membranes studied thus far in containing only 20% protein.

There is evidence suggesting that inorganic ions may be essential for the stability of some membranes (see Chapter 19). Systematic analytical data on ion content of membranes are not available.

LIPID COMPOSITION

As we pointed out in Chapter 12, the major part of the lipid of biological membranes consists of phospholipids, sphingolipids, and glycolipids. These

[1] In Gram-negative bacteria the cell envelope contains a third layer for which the name "outer membrane" is more appropriate. Its major constituent is lipopolysaccharide, but typical membrane lipids and proteins are also present in significant amounts.

Table 18-1. Analytical Protein and Lipid Content of Several Membranes[a]

	Percent of Dry Weight	
	Protein	Lipid
Myelin	18	79
Human erythrocyte	49	43
Bovine retinal rod	51	49
Mitochondria (outer membrane)	52	48
Mycoplasma laidlawii	58	37
Sarcoplasmic reticulum	67	33
Gram-positive bacteria	75	25
Mitochondria (inner membrane)	76	24

[a] Based on data from Guidotti (1972). Where the figures given do not add up to 100%, the remainder is listed as "carbohydrate". Such carbohydrate would normally not be in free form, but as a constitutive part of glycoproteins or glycolipids.

are all amphiphile molecules with two hydrocarbon tails, except that cardiolipin, which is sometimes present in significant amount, is a dimer of two such molecules, with a covalent link between the head groups. The hydrophilic head groups of these lipids vary considerably between different membranes (Table 12-1), but the hydrocarbon chains are similar. They are always long and contain both saturated and unsaturated hydrocarbons. There is an important difference between bacterial membranes and membranes of higher organism. The lipids in bacteria are exclusively of the two-tail amphiphile variety, whereas membranes of higher organisms may in addition contain as much as 25% of cholesterol.

Examination of the properties of phospholipids in Chapters 12 and 13 has shown that the nature of the hydrocarbon chains can have an important influence. If these chains are all saturated, bilayers with ordered hydrophobic cores are readily formed, whereas a mixture of saturated and unsaturated chains forms bilayers with hydrophobic cores that are liquidlike at temperatures above 0°C. Head group variation, on the other hand, was found to have only a minor effect on the properties of the lipid bilayer, except in liquid crystalline structures where contacts between head groups play an important role. It was also shown in Chapters 12 and 13 that cholesterol is readily incorporated into phospholipid bilayers, and that this incorporation leads to marked effects on local fluidity. Although these conclusions were based on experiments using phospholipids, it can be anticipated that sphingolipids and glycolipids behave similarly.

It is seen therefore that the principal lipids of all membranes would exhibit similar behavior in aqueous solution in the absence of protein. Whether alone or in mixtures, they would tend to form bilayers with liquid cores, and the bilayers would tend to form closed vesicles. The specificity observed in the types of head group associated with different membranes (Table 12-1) would not seem to be an important structural factor except insofar as it may be involved in lipid–protein interaction. The only truly distinctive feature of the lipid composition from this point of view would seem to lie in the presence or absence of cholesterol.

POLYPEPTIDE CHAINS

Many membrane proteins are virtually insoluble in aqueous media and their separation and characterization by the classical methods of protein chemistry is impossible. It is possible, however, to dissolve most membranes completely in aqueous sodium dodecyl sulfate[2] solutions at concentrations in excess of the cmc. Lipid molecules become incorporated in the detergent micelles under these conditions (page 108) and proteins are dissociated to their constituent polypeptide chains and converted to soluble micellar rodlike complexes containing a very high ratio of detergent to protein (page 139). If disulfide bonds are reduced, as is customarily done, covalently linked polypeptide chains will be separated, and each individual chain will bind the same amount of dodecyl sulfate (1.4 g per g of protein; see Fig. 15-5). The length of the rodlike particles will become approximately proportional to molecular weight. The polypeptide chains can then be separated from each other and from lipid-containing micelles by the sensitive method of polyacrylamide gel electrophoresis. The mobility of each complex in this method is a measure of its size, that is, a measure of the molecular weight of the constituent polypeptide chain. In this way a reproducible analysis for the polypeptide chain content of any membrane is readily obtained, as illustrated by Fig. 18-5. (It should be observed that glycopolypeptides with large carbohydrate moieties will have anomalous mobilities in this system because they form particles in which the polysaccharide chain will extend laterally from the protein–detergent rod with a consequent increase in Stokes radius. Highly acidic or highly basic proteins may also behave anomalously because their high intrinsic charge may influence the electrophoretic mobility directly or by altering the amount of bound detergent. One should in

[2] Other simple detergents can undoubtedly be used, but sodium dodecyl sulfate has become a standard reagent for this purpose.

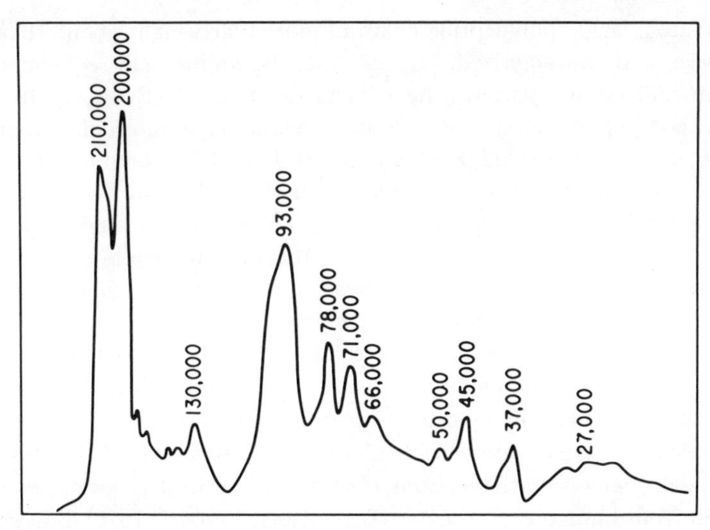

Fig. 18-5. Spectrophotometric scan of the polyacrylamide gel electrophoresis pattern obtained from the mixture of polypeptide chains of a human erythrocyte membrane, dissolved in aqueous sodium dodecyl sulfate. Molecular weight values corresponding to each peak are based on calibration with polypeptide chains of known molecular weight (taken from Trayer et al., 1971).

general regard molecular weights obtained by an empirical procedure of this kind as tentative until confirmed by more direct procedures.)

The results obtained by this procedure are in striking contrast to the conclusions based on the lipid compositions of membranes. Membranes are found to contain a diverse variety of polypeptide chains, and major components are often found to be uniquely associated with particular membranes.

Rod outer segments, for example, with a protein content of about 50%, have only a single major polypeptide chain constituent, with a molecular weight of about 37,000 (Cavanagh and Wald, 1969; Heitzmann, 1972; Robinson et al., 1972). This single polypeptide chain accounts for at least 80% of the total protein content of the membrane, and has been identified as the polypeptide chain of the visual receptor protein, rhodopsin. It is evident that the membranes of the rods serve the primary function of being a binding site for this protein.

Myelin membranes, which have an unusually low protein content, nevertheless contain a considerable variety of polypeptide chains. Several distinct polypeptide chains are seen in myelin from porcine brain, for example (Reynolds and Green, 1973). One of these is the so-called basic protein, which is water soluble, and associated with an experimentally inducible allergic disease, symptomatically resembling multiple sclerosis. This protein consists of a single polypeptide chain of molecular weight about 18,500 and has been well characterized, for example, the amino-acid sequences of the human and bovine varieties have been determined (Eylar, 1970). Other major polypeptide chains of porcine myelin have molecular weights of 52,000, 48,000, 35,000, 32,000, and 25,500. It may be noted that the myelin sheath is wrapped around the nerve axon (Fig. 18-3) in such a way as to permit the flow of aqueous solution between the myelin membranes to the axonal membrane. No transport across the myelin membrane is therefore required. Its function is not entirely clear, and no function can yet be assigned to any of the constituent polypeptide chains.

The erythrocyte membrane represents perhaps the simplest possible example of a cytoplasmic membrane since the erythrocyte has lost most of the normal functions of a living cell. Its protein content of about 50% represents a complex mixture of polypeptide chains, as illustrated by Fig. 18-5. Essentially identical results have been obtained in several laboratories for the protein from human erythrocytes (Berg, 1961; Lenard, 1970; Trayer et al., 1971). A minimum of 17 polypeptide chains differing in molecular weight is present. Forty percent of the protein is represented by two polypeptide chains of molecular weight 200,000 and 220,000. (The molecular weights have been confirmed by sedimentation equilibrium and other methods.) These large polypeptide chains may be unique constituents of erythrocyte membranes. Although polypeptide chains of similar molecular weight occur in other membranes, they are usually only minor constituents, and it has thus far not been demonstrated that they resemble the erythrocyte polypeptide chains in amino-acid composition or other properties. Another prominent peak in Fig. 18-5 occurs at an apparent molecular weight of about 100,000. This peak is asymmetric, indicating that it represents a mixture of polypeptide chains. Two distinct polypeptide chains have in fact been identified as components of this peak (Bretcher, 1971). One is a polypeptide chain of molecular weight about 105,000, and the other is a glycopolypeptide, containing more than 50% carbohydrate, which actually has a molecular weight much below the value of 90,000 to 100,000 indicated by its position on the gel pattern.

The complexity of polypeptide chain composition increases with the diversity of function of the membrane. Mitochondrial and bacterial membranes, for example, contain a greater variety of polypeptide chains than is shown in Fig. 18-5 for the erythrocyte membrane. Virus membranes, on the other hand, are generally very simple. Sindbis virus, for instance, has only two kinds of polypeptide chain in the entire virus particle (Strauss et al., 1968), one of which is associated with the viral membrane.

IDENTIFICATION AND CHARACTERIZATION OF SPECIFIC MEMBRANE PROTEINS

The information obtained from polypeptide chain analysis is limited to the size and relative abundance of the constituent chains. The chains are denatured in the detergent solution and their enzymatic or other biological activity is lost. Associations between chains to form multisubunit proteins are destroyed. Some additional information can be obtained. For example, glycopolypeptides can be identified on the gel by specific stains. In a few instances radioactive covalent labels can be attached in the intact membrane to identify a polypeptide chain as belonging to a particular enzyme. In general, however, the isolation and characterization of specific proteins requires the use of milder and more selective methods for disruption of the membrane than is afforded by the action of dodecyl sulfate. At the present time the number of proteins characterized by such procedures is small, and most of the polypeptide chains in a pattern such as that of Fig. 18-5 are unnamed and have unknown functions.

REFERENCES

Berg, H. C. (1969). *Biochim. Biophys. Acta*, **183,** 65.

Bretscher, M. S. (1971). *J. Mol. Biol.*, **58,** 775.

Cavanagh, H. E., and G. Wald. (1969). *Fed. Proc.*, **28,** 344.

Eylar, E. H. (1970). *Proc. Nat. Acad. Sci., U.S.A.*, **67,** 1425.

Guidotti, G. (1972). *Ann. Rev. Biochem.*, **41,** 731.

Heitzmann, H. (1972). *Nature New Biol.*, **235,** 114.

Heller, J., and T. Ostwald. (1972). *Ann. N.Y. Acad. Sci.*, **195,** 439.

Lenard, J. (1970). *Biochemistry*, **9,** 1129.

Reynolds, J. A., and H. O. Green. (1973). *J. Biol. Chem.*, **248,** 1207.

Robinson, W. E., A. Gordon-Walker, and D. Bownds. (1972). *Nature New Biol.*, **235,** 112.

Stoeckenius, W. (1969). In *Membranes of Mitochondria and Chloroplasts*, E. Racker, Ed., Van Nostrand-Reinhold Co., New York, Chapter 2.

Strauss, J. H., Jr., B. W. Pfefferkorn, and J. E. Darnell, Jr. (1968). *Proc. Nat. Acad. Sci. U.S.A.*, **59,** 533.

Trayer, H. R., Y. Nozaki, J. A. Reynolds, and C. Tanford. (1971). *J. Biol. Chem.*, **246,** 4485.

NINETEEN

THE PROBLEM OF MEMBRANE STRUCTURE

We have seen in the preceding chapter that the constituent lipids of biological membranes would in the absence of protein form vesicles bounded by a lipid bilayer. Cholesterol, if present, would be incorporated in the bilayer surface as illustrated schematically by Fig. 12-3. Thus membrane lipids by themselves would be expected to form a structure resembling the erythrocyte membrane shown in Fig. 18-1. Most membranes, however, contain only about 50% lipid by weight or even less (Table 18-1). It thus cannot be assumed that the constituent lipids will play the major role in structural organization. For membranes containing as much as 75% protein it would, in fact, be logical to expect that the protein content should dominate the structure.

The available data on the interaction between proteins and amphiphiles, presented in Chapter 15, reinforce this conclusion. It was shown there that complex formation between protein and amphiphile, if it involves the association of the amphiphile hydrocarbon chains with the protein surface, competes with self-association of the amphiphile to form micellar aggregates. Formation of protein–lipid complexes is therefore expected to withdraw some lipid hydrocarbon chains from the lipid pool capable of bilayer formation and to reduce the fraction of the membrane components that might spontaneously form bilayer vesicles to less than the already small lipid content given in Table 18-1. The major structural units of the membrane would reasonably be expected to be protein–lipid complexes.

Many models for membrane structure based on protein–lipid structural units have been proposed. It has been suggested, for example, that there is a unique structural protein, common to all membranes, that is capable of forming an underlying framework on which lipids and functional proteins could be deposited in an ordered arrangement. This suggestion, however, is incompatible with the compositional data of the previous chapter, which showed that no polypeptide chain can be identified as a common abundant constituent of all membranes. On the contrary, each type of membrane appears to contain a specific set of polypeptide chains, and it is probable that most of them are involved in the unique functional properties of the membrane. In the case of retinal rod outer segments, for example, the photoreceptor protein rhodopsin constitutes at least 80% of the total polypeptide chain content. Other proposals have visualized the typical membrane as an assembly of globular lipoprotein complexes.

The view that protein–lipid complexes are the intrinsic building blocks of membranes, plausible though it may be, is not supported by experimental data. The subject has been thoroughly reviewed by Stoeckenius and Engelman (1969), and they come to the conclusion that experimental data favor lipid bilayers as the predominant structural feature of most membranes. In spite of the high ratio of protein to lipid, the basic framework of cell membranes appears to be the kind of structure that the lipids alone would adopt if no protein were present at all. Some of the evidence, and the unanswered questions raised by it, will be considered below.

STRUCTURAL STUDIES

The idea of a lipid bilayer as a core of the typical membrane was first proposed by Gorter and Grendel (1925), who extracted the lipid from erythrocyte membranes, compressed it at a water–air interface, and found that it occupied a surface area equal to double the external area of the cells from which the lipid was derived. Danielli and Davson (1935) interpreted surface tension measurements (incorrectly, as it turns out) to indicate that lipid head groups were covered by layers of protein, thus accounting for both protein and lipid in terms of a bilayer model. With the advent of electron microscopy, the triple-layered structure shown in Fig. 18-1 was found to be an almost ubiquitous image of a membrane. Its similarity to the structures seen in electron micrographs of pure lipid liquid crystals (e.g., Fig. 12-1), the corroborative evidence from X-ray diffraction (see below), and the dimensional compatibility of the image with a bilayer structure, led Robertson, who was himself responsible for a major portion of the experimental work, to state the bilayer hypothesis in its most forceful terms (see review by Robertson, 1960). Since 1960, arguments have been raised against the evidence from electron microscopy, because samples have to be stained with heavy metals and dehydrated to obtain sufficient contrast between the sample and the grid upon which it is placed, but these arguments have been

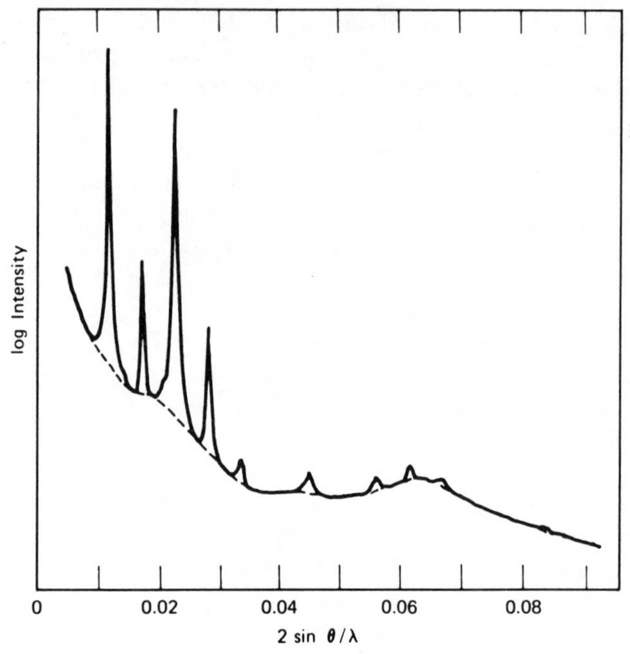

Fig. 19-1. (*a*) Scattered intensity (log scale) as a function of diffraction angle (θ) for myelin from rabbit sciatic nerves. λ is the wavelength of the X-rays.

rebutted fairly convincingly (Stoeckenius and Engelman, 1969; Robertson, 1972).

X-ray diffraction measurements are less open to question than electron micrographs, and it is worthwhile to describe the data from this source in some detail. X-ray studies of myelin (Fig. 18-3) and retinal rod outer segments (Fig. 18-4) are particularly interesting, for in these two systems nature provides us with a ready-made oriented stack of membranes that may be treated as a one-dimensional crystal. Diffraction from these systems yields several diffraction peaks (Fig. 19-1), which may be analyzed by the methods of X-ray crystallography to yield a low-resolution electron density profile in a direction perpendicular to the plane of the membranes. The phases of the individual peaks may be determined with reasonable certainty by swelling the stack of membranes and assuming that this alters the spacing between membranes, but not the dimensions of individual membranes.

The first X-ray studies of myelin were carried out by Schmitt et al. (1941). The results obtained were similar to those obtained by the same group for hydrated lipids (page 103) and demonstrated the bilayer structure of the mem-

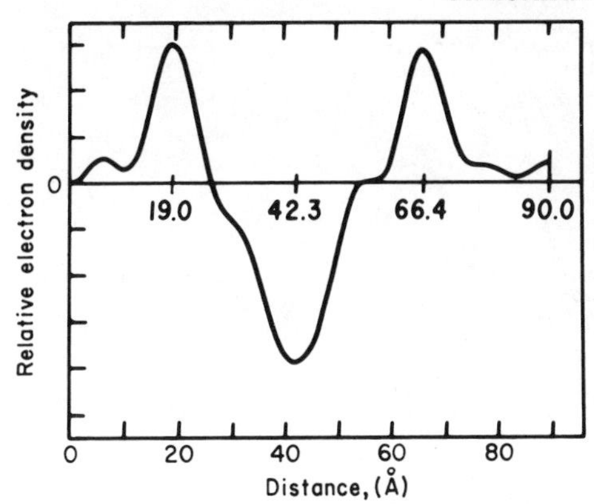

Fig. 19.1. (*continued*) (*b*) Electron density profile of a single membrane calculated from this diffraction pattern. The solvent density is taken as zero (taken from Caspar and Kirschner, 1971).

branes convincingly. A spacing of 4.7 Å was obtained in the direction parallel to the bilayer plane, a spacing now known to be characteristic of parallel hydrocarbon chains in bilayers with liquidlike cores. Studies with greater precision and better resolution have since been made in other laboratories (Finean and Burge, 1963; Worthington and Blaurock, 1969), culminating in the work of Caspar and Kirschner (1971), in which a resolution of 10 Å is attained. The resulting electron density profile is shown in Fig. 19-1, and is remarkably similar to that of cholesterol-containing phospholipid bilayers (Fig. 12-3) except that the cholesterol (26% of the total myelin lipid) is not symmetrically distributed.

Myelin membranes cannot be considered as characteristic of membranes in general because of their low protein content (Table 18-1) and especially as one of the major proteins, myelin basic protein, is easily removed and almost certainly located on an external surface of the membrane. The results obtained for retinal rod discs are therefore of greater interest, since the disc membranes contain 50% protein and, although most of the protein represents a single species, the photoreceptor protein rhodopsin, this protein is intimately associated with the membrane and cannot be removed unless the membrane is first disrupted with detergents. X-ray diffraction measurements of high precision have been made by Blaurock and Wilkins (1969) and by Corless (1972), with essentially identical results. The electron density profile of a single membrane is shown in Fig. 19-2. It consists of two peaks of high electron density, 40 Å apart, and a central trough of electron density below that of water. As previously indicated (page 105), this feature indicates that the center of the membrane is rich in terminal CH_3 groups of hydrocarbon chains. Retinal rod membranes contain little or no cholesterol, and

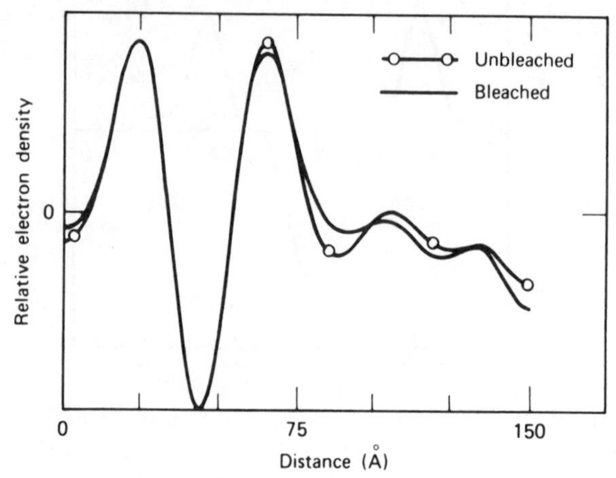

Fig. 19-2. Electron density profile for a single membrane from retinal rod outer segments. Essentially identical results are obtained regardless of whether the photoreceptor protein rhodopsin is bleached or unbleached. The solvent density is taken as zero (taken from Corless, 1972).

the shoulders in the electron density profile that are seen in myelin are not observed here.

Figure 19-2 provides unambiguous evidence that the predominant structural feature of the retinal rod disc membrane is a bilayer similar to that formed by phospholipid alone. The central trough is particularly significant as it can arise only from a region rich in terminal CH_3 groups. Model structures based on protein–lipid structural units appear to be rigorously excluded, since they could not give rise to this feature of the results.

The results leave one major problem unresolved. What is the location of the protein in the membrane? Blaurock and Wilkins (1969) made a careful analysis of the total amount of lipid per unit area in the photoreceptor membrane and concluded that it is insufficient for a pure lipid bilayer with the observed peak-to-peak distance: this distance would have been 22 to 28 Å instead of 40 Å if lipid alone had been responsible for the observed diffraction pattern. Blaurock and Wilkins initially accounted for this discrepancy by speculating that rhodopsin might be intimately associated with the lipid head groups, so that the electron density peaks in Fig. 19-2 would represent the center of an area containing both protein and lipid head groups, but they subsequently reinterpreted their data in terms of penetration of the bilayer by a substantial portion of the rhodopsin molecule, with the 40 Å separation representative of the separation between head groups alone (Blaurock and Wilkins, 1972). Possible evidence for the latter interpretation is provided by the observation of a diffuse reflection indicating a spacing of about 55 Å in the plane of the bilayer, which could represent the average distance between centers of *mobile* rhodopsin molecules, mobility being conferred on them by the fluidity of the bilayer. A similar spacing had been previously observed by Blasie et al. (1969) in a preparation of centrifuged discs and had been similarly interpreted. Blasie et al. found that the reflection corresponding to this spacing was sharpened when the membranes were first treated with antibody specific for rhodopsin, indicating that combination with the antibody reduces the mobility of the protein. This experiment also indicates that at least a part of the rhodopsin molecule must extend into the aqueous medium external to the membrane.

Wilkins et al. (1971) have carried out X-ray diffraction measurements on random dispersions of erythrocyte membranes and membranes from the microorganism *Mycoplasma laidlawii*. The data obtained from such dispersions are of course much inferior to data obtained from ordered arrays, since a continuous distribution of scattered intensity is obtained with only two or three broad peaks. Nevertheless, the patterns obtained (which were similar to diffraction patterns from dispersions of egg phosphatidyl choline) strongly indicated that a bilayer structure is the dominant one in these membranes also. *Mycoplasma laidlawii* is an organism requiring exogenous fatty acids, so that the hydrocarbon chain content of the constituent lipids can be controlled in the laboratory. When a relatively high proportion of saturated fatty acids is provided, the hydrocarbon chains may be ordered or disordered, depending on the temperature, as in pure phospholipid bilayers. Temperature-dependent changes in the X-ray diffraction pattern, consistent with those that occur in pure phospholipid systems, have been observed by Engelman (1971).

X-ray diffraction measurements have also demonstrated that membranes present in certain viruses have a bilayer-dominated structure. Sindbis virus is an example. This is a spherical virus bounded by a membrane and containing a glycoprotein external to the membrane. The overall radius is 350 Å, but the outer shell occupied by the glycoprotein is penetrable by solvent. X-ray diffraction measurements of suspensions of these viruses permit evaluation of an electron density distribution function (spherical symmetry assumed) with a resolution of 28 Å (Harrison et al., 1971a). The result obtained is shown in Fig. 19–3. The deep minimum at a radial distance of 232 Å is virtually conclusive evidence for the location of the center of a lipid bilayer at that position. Similar results have been obtained with bacteriophage PM2, another membrane-containing virus (Harrison et al., 1971b).

FLUIDITY AND MOTILITY

The apparent similarity between biological membranes and pure lipid bilayer systems extends to measurements of the type considered in Chapter

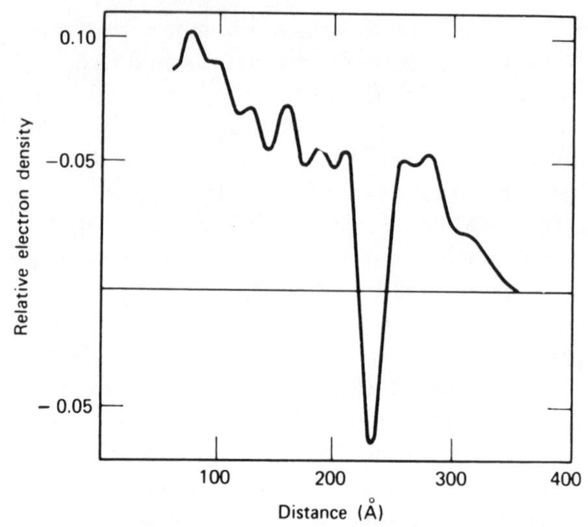

Fig. 19-3. The spherically averaged electron density of whole Sindbis virus, as a function of radial distance from the center. The solvent electron density is taken as zero (taken from Harrison et al., 1971a).

13, which provide direct information about the fluidity of the membrane core and the motility of lipid molecules within it. Hubbell and McConnell (1969) have shown, for example, that spin labels inserted into axonal membranes from the walking-leg nerve of the lobster display rapid anisotropic motion similar to that observed when these labels are inserted into phospholipid vesicles, and have observed a similar increase in local fluidity as the label is moved along the hydrocarbon chain from the head group toward the terminal methyl group. Erythrocyte membranes were found in the same study to have a much less fluid hydrophobic core, but the hydrocarbon chains were found to be oriented perpendicular to the membrane surface and fluidity was found to increase as the spin label was moved closer to the terminal methyl group. The results appear to indicate that the erythrocyte lipids are in a bilayer arrangement, but that the motion of their hydrocarbon chains is more restricted than in comparable pure lipid systems or in the lobster axonal membrane. Restrictions on the mobility of hydrocarbon chains in erythrocyte membranes have also been observed by nmr measurements (Chapman et al., 1969). In another important paper, Scandella et al. (1972) have shown that spin-labeled phospholipid molecules are capable of lateral diffusion in sarcoplasmic reticulum membranes, and that the rate of diffusion is about as fast as in a pure phospholipid bilayer.

As was mentioned before, the organism *Mycoplasma laidlawii* is viable with a high content of saturated hydrocarbon acyl chains attached to the lipid molecules, and the transition from ordered to liquidlike bilayer core can therefore be observed in *Mycoplasma* membrane preparations. Steim et al. (1969) have studied the transition using differential thermal calorimetry (page 113) and found that the transition occurs at the same temperature in the membrane as in protein-free dispersions of the lipids alone. Moreover, the lowering in the transition temperature upon substitution of unsaturated for saturated fatty acid chains was the same in both systems. In the experiments using whole membranes, a second transition corresponding to thermal denaturation of the membrane proteins was observed. The temperature of this transition was found to be independent of the fatty acid composition. On the basis of these experiments the membrane proteins and the hydrocarbon chains of the lipids would seem to be quite independent and unrelated components of the membrane. However, the total heat absorbed in the lipid transition per mole of lipid is less in the whole membrane than in the pure lipid dispersion by about 25%, and this may indicate that some fraction of the lipid in the whole membrane does not participate in the transition, perhaps because it is held to the membrane protein in some conformation quite different from that of a bilayer, and not subject to change with temperature.

It may be noted in this connection that Lenard and Singer (1968) have found that only 70% of the phospholipid head groups of erythrocyte membranes are susceptible to the action of phospholipase C, an enzyme that hydrolyses phospholipids to diacyl glycerides and water-soluble phosphorylated head groups. In a subsequent paper from the same laboratory (Glaser et al., 1970) it is suggested that the phenomenon of having a part of the membrane lipids withdrawn from a bilayer arrangement by interaction with membrane proteins may be a general one.

LOCATION OF MEMBRANE PROTEINS

The structural similarities between biological membranes and pure lipid bilayers revealed by the physical methods discussed above are, as stated at the beginning of this chapter, unexpected, and the conclusions drawn from the measurements must be incomplete. Membranes do differ from pure lipid systems, and these differences must have an explanation at the structural level. It is necessary (1) to account for the permeability of cell membranes to a variety of substances to which lipid bilayers are impermeable or much less permeable, and (2) to find a place in the structure for the proteins, which in most cases comprise 50% or more of the total mass of the membrane. The two questions raised are interrelated since the specific physiological properties of a membrane are presumably conferred upon it by the constituent proteins.

As noted in Chapter 16, there are virtually no data on the interaction between membrane proteins and specific lipids. There is some information on the conditions required to remove membrane proteins from a membrane, and this indicates that some proteins are bound to external surfaces as visualized by the early schematic membrane models of Danielli and Davson (1935) and Robertson (1960). Proteins assigned to this category are proteins that can be removed from membranes by mild reagents without effect on the integrity of the membrane per se, and the basic protein of myelin was mentioned earlier as an example. The majority of membrane proteins, however, do not fall into this category and can be removed from the membrane only by the use of detergents, which at the same time disrupt the bilayer structure. Such proteins must interact more intimately with the membrane lipids and presumably penetrate into the bilayer region. Many proteins in this category may lie partly in and partly out of the bilayer, as was suggested above for the retinal rod membrane protein, rhodopsin.

The most direct evidence on this subject comes from experiments using freeze fracture electron microscopy. In this technique a specimen in aqueous suspension is frozen and then fractured by a blow from a microtome knife. The fracture will tend to seek out planes containing the weakest intermolecular bonds which, in a system containing lipid bilayers, would be the central region of the bilayer in which the ends of the lipid hydrocarbon chains are concentrated.[1] The interior of the bilayer is thereby exposed and a replica of its morphology can be obtained by evaporation of a film of metal onto it. When these replicas are examined in the electron microscope, smooth fracture planes are obtained from pure lipid systems and also from myelin (Branton, 1967). Most other membranes produce rough surfaces suggesting that the fracture has occurred around particulate obstacles. Convex surfaces (connected to the cell interior) and concave surfaces (connected to the external surface of the cell) generally are quite distinct. Sometimes complementary surfaces are seen, with depressions in one surface matching protuberances in the other. Data of this kind have been obtained for retinal rods (Clark and Branton, 1968), mitochondria (Wrigglesworth et al., 1970), erythrocyte membranes (Fig. 19-4) and other membranes. For erythrocytes it has been unequivocally established (Pinto da Silva and Branton, 1970; Tillack and Marchesi, 1970) that the fracture faces display the inside and not the exterior surfaces of the membrane. Although it cannot be established

[1] Fracture of the frozen specimen is performed in a vacuum, so that freshly exposed surfaces consist of *unbonded* molecular surfaces. The situation is completely different from disruption of a bilayer in an aqueous medium, where freshly created surfaces would be exposed to water, so that the hydrophobic areas would be the least likely to be exposed.

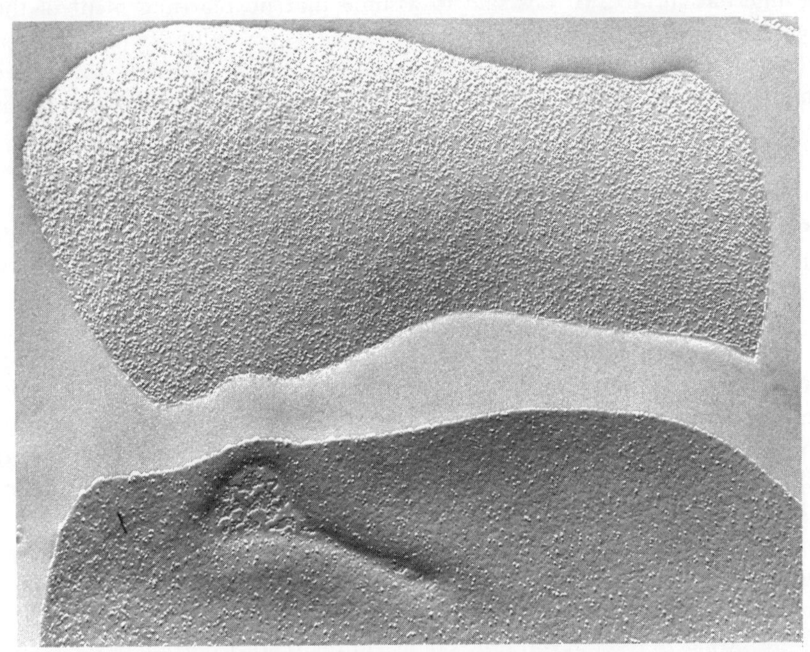

Fig. 19-4. Freeze fracture electron micrograph of an erythrocyte membrane. The upper part of the photograph represents the side of the membrane facing the cell interior and contains many more protuberances than the lower part of the photograph, which is the out-facing side of the membrane of an adjacent cell. (Obtained through the courtesy of Knute A. Fisher, Department of Botany, University of California, Berkeley).

that the particles observed in these pictures represent actual outlines of protein molecules or clusters of protein molecules, they do represent severe distortions of the bilayer interior for which proteins penetrating into the bilayer must be directly or indirectly responsible.

Experiments that have been done to probe accessibility of membrane proteins to reagents added from the outside are of some interest in relation to the question of where these proteins are located. They indicate that proteins are generally not symmetrically distributed between the two sides of a membrane, as is to be expected on the basis of a membrane's functional role. (The same phenomenon is observed in freeze fracture electron micrographs, as seen from Fig. 19-4). In erythrocytes, for example, only two of the large number of constituent polypeptide chains (Fig. 18-5) are accessible to reagents added to the outside surface of the intact cell (Berg, 1969; Phillips and Morrison, 1971; Bretscher, 1971a). All of the proteins can however react with suitable reagents after the cell has been lysed to expose the internal surface. If it is safe to assume that no rearrangement of the organization of the membrane occurs during this procedure (which has not been established), these experiments imply that no polypeptide chains are wholly embedded in the membrane, but that they all have exposed portions on one side or the other. It is of interest to note that Bretscher (1971b, c) has found that the two erythrocyte proteins that are exposed on the external surface of the intact cell are labeled in different positions after lysis. With the foregoing proviso that membrane organization remains intact, this experiment suggests that these two proteins may traverse the membrane from one side to the other.

If we consider these observations in combination with the structural data for two membrane proteins that were discussed in Chapter 16 we can see a way out of the apparent contradiction between a lipid-dominated structure for membranes and the high protein content. It was shown in Chapter 16 that a segment of polypeptide chain containing only 23 amino acids would suffice to bind the major glycoprotein of erythrocytes to the membrane. (This is one of the two proteins believed to traverse the membrane.) Similarly, evidence was presented that a segment of no more than 44 amino acids is involved in the association of cytochrome b_5 with microsomal membranes. In each example the remainder of the protein molecule might project into the adjacent fluid beyond the bounds of the bilayer portion of the membrane. In other words, if these results are generally applicable, the amount of protein actually within the domain of the bilayer structure could easily be far less than the *analytical* protein content.

THE FLUID MOSAIC MODEL

Both the structural information cited above and functional considerations—that is, the permeability of cell membranes to specific metabolites—have led to the suggestion that typical membranes may have a mosaic structure. What is meant by this is that membranes are visualized as possessing a mixture of protein-rich regions responsible for the functional properties of the membrane and essentially pure lipid regions that serve to maintain the integrity and cohesiveness of the structure as a whole. This general idea has been held by many workers for a long time (e.g., Collander and Bärlund, 1933; other early references are given by Höber, 1945) and has most recently been strongly advocated by Singer and Nicolson (1972). By virtue of the presence of unsaturated acyl chains, the bilayer part of a mosaic structure will generally be fluid, and the protein-rich portions will consist of "islands" in a fluid bed. Although these portions may themselves be quite rigidly structured, they, too, will be mobile in the membrane, as a result of the flow of lipid around them. In the case of proteins such as cytochrome b_5, that have their functional parts in the cytoplasm outside the bilayer, the "islands" may be quite small, consisting of a short segment of polypeptide chain and presumably some lipid molecules that are involved in anchoring the protein to the membrane. Proteins involved in transport, on the other hand, may form much larger "islands" that may contain parts of several polypeptide chains, and could easily include water-filled channels.

There is experimental support for the mobility of proteins in membranes. Evidence for rotational freedom comes, for example, from observations on the dichroism associated with absorption of polarized light by the photoreceptor protein rhodopsin. The chromophore of this protein, in the intact membrane, has been shown by these studies to be in a state of rapid rotational motion (Brown, 1972; Cone, 1972). If the membrane-bound protein is immobilized by cross-linking, the rotational motion is eliminated, indicating that it is the protein molecule itself, or a major part of it, that is responsible for the mobility in the native state.

Evidence for translational mobility is provided by the work of Frye and Edidin (1970) who fused cells from tissue culture lines of mouse and human origin, each bearing surface proteins with distinct antigenic determinants. The movement of the antigenic sites could be followed by use of fluorescent antibodies, and it was observed that both antigens became randomly distributed over the fused hybrid cells within about 40 minutes. Another example is provided by the microsomal protein cytochrome b_5, already mentioned above. This protein can be combined with vesicular preparations of microsomal lipids, and the product obtained in this way cannot be distinguished from endogenous membrane-bound protein. The amount of protein in the vesicular preparation can be varied over a wide range, and the effect of altering the amount on the rate of reaction with the specific enzyme, cytochrome b_5 reductase, strongly suggests that both cytochrome b_5 and the reductase are mobile in the membrane and react with each other by random collision (Strittmatter et al., 1972). Evidence that rhodopsin has translational mobility in photoreceptor membranes was given earlier, and several additional examples are cited by Singer and Nicolson (1972).

This concept of mobility of the protein-rich regions of the membrane, coupled with the probability that the amount of protein actually within the domain of the bilayer structure is considerably less than the analytical protein content, may account for the inability to detect electron-dense regions corresponding to membrane proteins in most X-ray diffraction studies. In the fluid mosaic model the protein-rich regions would contribute electron density to both the hydrophobic interior of the membrane and the area occupied by the head groups, as well as to the aqueous medium between neighboring membranes in the stacked arrays. Because of the mobility this electron density would be expected to be evenly distributed along the plane of the membrane, and it would thus constitute a more or less constant background above which the ordered electron density pattern contributed by the pure bilayer regions of the mosaic structure would be seen. Considering the low resolution of the X-ray diffraction data, it is plausible that the resulting electron density profile could not be readily distinguished from that observed in a pure lipid system. The presence of protein would become apparent only through a discrepancy between the analytical lipid content and the total mass required to fill the space occupied by the bilayer, such as was observed for retinal rod membranes by Blaurock and Wilkins (1969, 1972).

The observed fluidity of membrane lipids would also not be inconsistent with a model of this kind, since spin labels would presumably be predominantly located in the lipid bilayer part of the structure, and the experimental data would not provide information on whether this part of the structure represents all or only part of the membrane. As noted above, the heat absorbed in the transition from ordered to fluid arrangement of the hydrocarbon chains in *Mycoplasma* membranes suggests that about 25% of the constituent lipid is not involved in the transition.

RESERVATIONS AND CONCLUSIONS

The view of a biological membrane that has emerged in the preceding portions of this chapter was reached on the basis of only a handful of structural studies, and great caution must be exercised in accepting it even as a working hypothesis. It should be noted in particular that the evidence for a mosaic type of membrane with a fluid lipid bilayer as framework is strongest for membranes that are not of the most basic type, that is, not membranes surrounding a living cell and controlling transport of metabolites in and out of it. New structural principles may well emerge from detailed examination of other membranes, especially the very protein-rich membranes of mitochondria and bacteria. In addition, there are many experimental observations that are inconsistent with a fluid mosaic model in its simplest form, as is illustrated by the following examples:

1. *Asymmetric lipid distribution.* We have seen in Chapter 13 that lipid molecules in a pure lipid bilayer are capable of rapid anisotropic rotational motion and of rapid lateral diffusion in the plane of the bilayer. Exchange of lipids between the two membrane faces occurs much more slowly, but the half time for this process in phosphatidylcholine vesicles (6.5 hr at 30° C, see page 118) is sufficiently small to ensure equal distribution of lipids on the two sides of a mature membrane, except insofar as the distribution may be influenced by affinity of proteins, localized on one side or the other, for specific lipids. It is therefore surprising to find that cholesterol

in myelin membranes is asymmetrically distributed, as seen in Fig. 19-1*b*. Cholesterol has only a weakly polar head group, and the hydrophobic core of the membrane should be less of a barrier to it than it is to phosphatidycholine. Moreover, myelin contains less protein than any other natural membranes, and displays no protuberances in freeze fracture electron micrographs, so that interference with a symmetric distribution from that source should be minimal.

A similar problem is posed by the evidence of Bretscher (1972), suggesting that phospholipids of erythrocyte membranes are asymmetrically distributed, choline head groups of phosphatidylcholine and sphingomyelin being almost exclusively at the external cell surface, whereas phosphatidylserine and phosphatidylethanolamine face the interior of the cell. An entirely different problem, apparently *too rapid* exchange of lipids, arises from the rapid growth and multiplication of microorganisms. Synthesis of phospholipids must occur at the cytoplasmic side of the membrane, but newly synthesized lipid molecules find their way to the outer surface in a time of the order of minutes.

2. *Role of Inorganic Cations.* There are some observations that indicate that cations can play an important role in membrane structure (Reynolds, 1972). The most important of these concern the erythrocyte membrane. The intact erythrocyte can be swollen and rendered "leaky" by exposure to a hypotonic aqueous medium. Hemoglobin and other cytoplasmic constituents escape when this is done, but the membrane remains intact by morphological criteria and retains most of its normal enzymatic function. The membrane, depleted of its contents, is rapidly broken up if inorganic cations are removed by dialysis against water or by organic sequestering agents, yielding a mixture of membrane fragments, protein-lipid complexes and lipid-free water-soluble proteins. The original structure is restored if inorganic cations are added back to the medium. These experiments suggest the possibility that the erythrocyte membrane may not be based on a continuous closed bilayer framework, but that it may consist instead of discontinuous bilayer patches, perhaps bounded by protein, with cation bridges joining patches to one another. It is not known at the present time how general this phenomenon might be, because even simple analytical data for the content of inorganic ions of membranes are usually unavailable.

Even if these and similar phenomena can be satisfactorily explained and the mosaic model suggested earlier proves to be generally valid, it could not be regarded as a "solution" to the problem of membrane structure. In earlier discussions in this book, dealing with micelle formation, bilayer structure in pure lipid systems, and protein–amphiphile interaction, the intermolecular organization could be characterized in considerable detail, and the forces responsible for the observed structures and properties could be explicitly defined. Explicit knowledge of this kind is not available in regard to the structure of biological membranes. We do not know what kind of interactions hold the protein in place in the membrane. It may be assumed that lipids are involved, but the relative importance of the hydrocarbon moieties and polar head groups has not been established. Moreover, assuming the validity of the mosaic model, we do not know how the protein-bound lipids merge with the free lipid molecules in the bilayer regions. As far as the protein intrusions into the membrane are concerned, we do not know in any given instance whether they involve single polypeptide chains or complex clusters. We cannot yet identify or characterize structural regions that may be associated with the transport of ions or polar molecules across membrane.

Thus the problem of membrane structure, at the time at which this is written, is still very much an unsolved problem, and the object of intensive research. It is likely that further progress on this problem cannot come from direct examination of membranes alone, but that this will have to be supplemented by detailed characterization of membrane proteins and the manner in which they interact with membrane lipids.

REFERENCES

Berg, H. C. (1969). *Biochim. Biophys. Acta*, **183,** 65.
Blasie, J. K., C. R. Worthington, and M. M. Dewey. (1969). *J. Mol. Biol.*, **39,** 407.
Blaurock, A. E., and M. H. F. Wilkins. (1969). *Nature*, **223,** 906.
Blaurock, A. E., and M. H. F. Wilkins. (1972). *Nature*, **236,** 313.
Branton, D. (1967). *Exp. Cell. Res.*, **45,** 703.
Bretscher, M. S. (1971a). *J. Mol. Biol.*, **58,** 775.
Bretscher, M. S. (1971b). *J. Mol. Biol.*, **59,** 351.
Bretscher, M. S. (1971c). *Nature New Biol.*, **231,** 225.
Bretscher, M. S. (1972). *Nature New Biol.*, **236,** 11.
Brown, P. K. (1972). *Nature New Biol.*, **236,** 35.
Caspar, D. L. D., and D. A. Kirschner. (1971). *Nature New Biol.*, **231,** 46.
Chapman, D., R. B. Leslie, R. Hirz, and A. M. Scanu. (1969). *Biochim. Biophys. Acta*, **176,** 524.
Clark, A. W., and D. Branton. (1968). *Z. Zellforsch. Mikrosk. Anat.*, **91,** 586.
Collander, R., and H. Bärlund. (1933). *Acta Bot. Fenn.*, **11,** 1.
Cone, R. A. (1972). *Nature New Biol.*, **236,** 39.
Corless, J. M. (1972). *Nature*, **237,** 229.
Danielli, J. F., and H. Davson. (1935). *J. Cell. Comp. Physiol.*, **5,** 495.
Engelman, D. M. (1971). *J. Mol. Biol.*, **58,** 153.
Finean, J. B., and R. E. Burge. (1963). *J. Mol. Biol.*, **7,** 672.
Frye, L. D., and M. Edidin. (1970). *J. Cell Sci.*, **7,** 319.
Glaser, M., H. Simpkins, S. J. Singer, M. Sheetz, and S. I. Chan. (1970). *Proc. Nat. Acad. Sci. U.S.A.*, **65,** 721.
Gorter, E., and F. Grendel. (1925). *J. Exp. Med.*, **41,** 439.
Harrison, S. C., A. David, J. Jumblatt, and J. E. Darnell. (1971a). *J. Mol. Biol.*, **60,** 523.
Harrison, S. C., D. L. D. Caspar, R. D. Camerini-Otero, and R. M. Franklin. (1971b). *Nature New Biol.*, **229,** 197.
Höber, R. (1945). *Physical Chemistry of Cells and Tissues*, The Blakiston Co., Philadelphia, Pa.
Hubbell, W. L., and H. M. McConnell. (1969). *Proc. Nat. Acad. Sci. U.S.A.*, **63,** 16; **64,** 20.
Lenard, J., and S. J. Singer. (1968). *Science*, **159,** 738.

Phillips, D. R., and M. Morrison. (1971). *Biochemistry*, **10,** 1766.
Pinto da Silva, P., and D. Branton. (1970). *J. Cell Biol.*, **45,** 598.
Reynolds, J. A. (1972). *Ann. N.Y. Acad. Sci.*, **195,** 75.
Robertson, J. D. (1960). *Progr. Biophys, Biophys. Chem.*, **10,** 344.
Robertson, J. D. (1972). *Arch. Intern. Med.*, **129,** 202.
Scandella, C. J., P. Devaux, and H. M. McConnell. (1972). *Proc. Nat. Acad. Sci.*, **69,** 2056.
Schmitt, F. O., R. S. Bear, and K. J. Palmer. (1941). *J. Cell Comp. Physiol.*, **18,** 31.
Singer, S. J., and G. L. Nicolson. (1972). *Science*, **175,** 720.
Steim, J. M., O. J. Edner, and F. G. Bargoot. (1968). *Science*, **162,** 909.
Stoeckenius, W., and D. M. Engelman. (1969). *J. Cell Biol.*, **42,** 613.
Strittmatter, P., M. J. Rogers, and L. Spatz. (1972). *J. Biol. Chem.*, **247,** 7188.
Tillack, T. W., and V. T. Marchesi. (1970). *J. Cell Biol.*, **45,** 649.
Wilkins, M. H. F., A. E. Blaurock, and D. M. Engelman. (1971). *Nature New Biol.*, **230,** 72.
Worthington, C. R., and A. E. Blaurock. (1969). *Biophys. J.*, **9,** 970.
Wrigglesworth, J. M., L. Packer, and D. Branton. (1970). *Biochim. Biophys. Acta*, **205,** 125.

AUTHOR INDEX

Numbers in parentheses refer to pages where references are listed.

SUBJECT INDEX